group 4	group 5	group 6	group 7	group 8
				2 Helium He 4.00
6 Carbon C 12.01	7 Nitrogen N 14.01	8 Oxygen O 16.00	9 Fluorine F 19.00	10 Neon Ne 20.18
14 Silicon Si 28.09	15 Phosphorus P 30.97	16 Sulphur S 32.06	17 Chlorine Cl 35.45	18 Argon Ar 39.45
32 Germanium Ge 72.59	33 Arsenic As 74.92	34 Selenium Se 78.96	35 Bromine Br 79.90	36 Krypton Kr 83.80
50 Tin Sn 118.69	51 Antimony Sb 121.75	52 Tellurium Te 127.60	53 Iodine I 126.90	54 Xenon Xe 131.30
82 Lead Pb 207.2	83 Bismuth Bi 208.98	84 Polonium Po (209)	85 Astatine At (210)	86 Radon Rn (222)

PERIODIC TABLE OF ELEMENTS

26 Iron Fe 55.85	27 Cobalt Co 58.99	28 Nickel Ni 58.70	29 Copper Cu 63.55	30 Zinc Zn 65.38
44 Ruthenium Ru 101.07	45 Rhodium Rh 102.91	46 Palladium Pd	47 Silver Ag	48 Cadmium Cd
76 Osmium Os 190.2	77 Iridium Ir 192.22			

ARTHUR GODMAN

BARNES & NOBLE THESAURUS OF SCIENCE

all fields of
scientific language
explained and illustrated

BARNES & NOBLE BOOKS
A DIVISION OF HARPER & ROW, PUBLISHERS
New York, Cambridge, Philadelphia, San Francisco
London, Mexico City, São Paulo, Sydney

Library of Congress Cataloging in Publication Data

Godman, Arthur.
    The Barnes & Noble thesaurus of science.

    Includes index.
    1. Science—Dictionaries.   I. Title.
Q123.G59   1983   503'.21   82-48830
ISBN 0-06-015176-5
ISBN 0-06-463580-5 (pbk.)
Printed in Spain by Heraclio Fournier SA

Contents

How to use this book

This book contains some 1500 basic scientific words divided into three main groups: physics, chemistry and biology. Some words may appear in two of the groups, e.g. motor[1] in physics, motor[2] in biology. Related words from other areas of science are also included: the physics part contains words from meteorology, chemistry contains words from geology and health and hygiene are found in biology. In addition the entries are arranged within the three main groups according to meaning, to help the reader to obtain a broad understanding of a particular area of science.

The main groups are further divided into subjects and parts of subjects. At the top of each page the subject is shown in bold type and the part of the subject in lighter type, for example on pages 118 and 119:

118 · CHEMICAL REACTIONS/ACIDS, BASES, SALTS

GEOLOGY/MINERALS · **119**

---

1. To find the meaning of a word

Look for the word in the alphabetical index at the end of the book, then turn to the page number listed.

The description of the word may contain some words with arrows in brackets after them. This is an indication that these words with arrows are defined nearby, and looking at that word may help in understanding the word being looked up:

(↑) the related word appears above or on the opposite page

(↓) the related word appears below or on the opposite page

(p.21) the related word appears on page 21

---

2. To find related words

Look for the word you are starting from in the index and turn to the page number shown. Because this book is arranged by ideas, all the related words are found in a set on that or the opposite page. The illustrations will also help, as it is often difficult to give a clear description using only words.

For example, the important reproductive parts of a flower are all defined on page 211. They include *stamen*, *anther*, *stigma*, *style*, *ovary*, and *carpel*. *Pollen* and *pollination* are also described on that page, as are all the words necessary to understand pollination.

3. To review a subject

There are two methods. Firstly, you may wish to see if you know all the words used in that subject. Secondly, you may wish to review your knowledge of a subject.

(a) To review the words used in the subject of *chemical solutions*. Look up 'SOLUTION' in the alphabetical index. Turn to the page indicated, page 89. There you find the following words: *dissolve, solution, solute, solvent, saturated, unsaturated*.

Turn over to page 90, and you will find more words of related meaning to 'SOLUTION'. They are: *insoluble, solubility, concentration, dilute*. On page 91 you will see the words: *miscible, immiscible, separate, precipitation*.

All these words are relevant when you wish to write about *chemical solutions*.

(b) To review your knowledge about a subject, for example *winds*. The only word that you can remember well is 'SEA-BREEZE'.

(i) Look up the alphabetical index and find 'SEA-BREEZE'. The page reference is 46.

(ii) Now turn to page 46. There you will find the words: *thermal, sea-breeze, land-breeze, trade wind, monsoon*.

The different kinds of winds are described. The explanation of each kind of wind depends on knowing the cause of a *thermal*, the first word on the list. After understanding this word, helped by a diagram, the other words become easier to understand. On looking at page 47 these words are seen: *whirlwind, typhoon, wind scale, cloud, cirrus, cumulus, nimbus, stratus*.

These words complete the subject of *winds*, and then follows information about *clouds*. So, by using pages 46 and 47, the subject of *winds* can be reviewed and made clearer still by referring to the illustrations.

4. To find a word to fit a required meaning

It is almost impossible to find a word to fit a meaning in most dictionaries. It is easy with this book. For example, if you have forgotten the name of the part of the eye that is receptive to light, all you have to do is find the page reference for eye in the index and turn to that page. There is a diagram of the eye so you can see the relation of the parts, and the parts of the eye are described there, as well. You can then pick out *retina* from the others.

Useful information in science

THE SI (SYSTEME INTERNATIONAL) UNITS

PREFIXES

PREFIX	FACTOR	SIGN	PREFIX	FACTOR	SIGN
milli-	$\times 10^{-3}$	m	kilo-	$\times 10^{3}$	k
micro-	$\times 10^{-6}$	μ	mega-	$\times 10^{6}$	M
nano-	$\times 10^{-9}$	n	giga-	$\times 10^{9}$	G
pico-	$\times 10^{-12}$	p	tera-	$\times 10^{12}$	T

BASIC UNITS

UNIT	SYMBOL	MEASUREMENT
metre	m	length
kilogramme	kg	mass
second	s	time
ampere	A	electric current
kelvin	K	temperature
mole	mol	amount of substance

DERIVED UNITS

UNIT	SYMBOL	MEASUREMENT
newton	N	force
joule	J	energy, work
hertz	Hz	frequency
pascal	Pa	pressure
coulomb	C	quantity of electric charge
volt	V	electrical potential
ohm	Ω	electrical resistance

COMMON ABBREVIATIONS

a.c.	alternating current
b.p.	boiling point
d.c.	direct current
e.m.f.	electromotive force
f.p.	freezing point
i.r.	infra red
m.p.	melting point
p.d.	potential difference
r.h.	relative humidity
s.t.p.	standard temperature and pressure
u.v.	ultra violet
v.p.	vapour pressure

COMMON CONSTANTS

s.t.p.   standard temperature and pressure, expressed as 1.00 atm or 760 mm Hg or 101 kPa ($= kNm^{-2}$) (Pa = pascal)

Standard volume of a mole of gas at s.t.p., $22.4 dm^3$

The Faraday constant, F, 96 500 $C mol^{-1}$

The Avogadro constant, L, $6.02 \times 10^{23} mol^{-1}$

Speed of light, c, $3.00 \times 10^8 ms^{-1}$

1 calorie = 4.18 J

Specific heat capacity of water, 4.18 $Jg^{-1} K^{-1}$

$\pi = \frac{22}{7}$ or 3.1416

Speed of sound in air, $3.3 \times 10^2 ms^{-1}$

Acceleration due to gravity, G, $9.81 ms^{-2}$

MATHEMATICAL SYMBOLS

$\equiv$	identically equal to
$\approx$	approximately equal to
$\propto$	varies directly as
∞	infinity
ab, a·b, a $\times$ b	a multiplied by b
a/b, $\frac{a}{b}$, ab^{-1}	a divided by b
sine (sin) cosine (cos) tangent (tan)	these are ratios which measure angles; they can be read from tables of angles

Physics

device (*n*) an object made for a special purpose to help us in our work.

tool (*n*) a device which is used with our hands to make furniture, machines, to build houses, to dig the ground.

instrument (*n*) a device which is used to do skilful work, or to measure accurately (↓).

dimension (*n*) the dimensions of a solid are its length, its breadth, and its height. Liquids and gases do not have dimensions. A solid has three dimensions; a flat surface has two dimensions (length and breadth) and a line has one dimension (length).

circumference (*n*) the line which forms a circle or is the limit of any area.

radius (*n*) the distance between the centre of a circle and its circumference or the centre of a sphere and its surface, *see diagram*.

diameter (*n*) a straight line passing from side to side through the centre of a circle or sphere, or the length of that line.

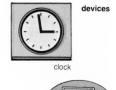

devices

clock

telephone

compass

tools

hammer

spanner

screwdriver

instruments

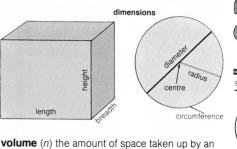

dimensions

height

length

breadth

diameter

radius

centre

circumference

volume (*n*) the amount of space taken up by an object in three dimensions.

measurement (*n*) the result of measuring or the action of measuring, e.g. (1) if a rope is measured and it is 2.5 m long, then its measurement is 2.5 m; (2) a ruler was used for the measurement of the rope.

accurate (*adj*) describes a measurement that is free from mistakes, e.g. using one instrument, an object's length is 26 mm. Using another instrument, the object's length is 26.2 mm; this is a more accurate measurement. **accuracy** (*n*).

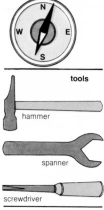

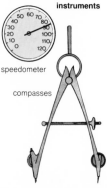

speedometer

compasses

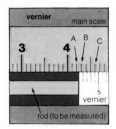

The 0 mark of the vernier is between lines A and B. At C a unit of the main scale meets the unit 4 of the vernier scale exactly showing that the rod is 4.24 cm or 42.4 mm.

scale (*n*) a set of marks, with numbers, rising from a low value to a high value.

vernier (*n*) part of an instrument used for measuring accurately, e.g. length to 0.1 mm. The vernier moves along a main scale. In the diagram the vernier reads 4.24 cm or 42.4 mm.

travel (*v*) to go from one place to another place; to go along a path, e.g. the Earth travels round the sun; a motor-car travels from one town to another town.

motion (*n*) movement; the action of travelling; going from one place to another place, e.g. the Earth is in motion round the sun; a motor-car is in motion when it is travelling. Motion is either in a straight line or in a curved line.

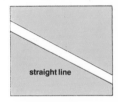
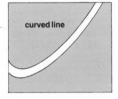

speed (*n*) the distance travelled divided by the time taken to travel the distance. If a motor-car travels 35 km in half an hour, its speed is 70 km per hour. The motion of the motor-car can be along a straight line or along a curved line.

velocity (*n*) speed in a certain direction. A motor-car travelling along a straight road with a speed of 70 km/hour also has a velocity of 70 km/hour. A motor-car travelling at 70 km/hour round a bend in a road has a velocity that is continuously changing as the road is not in a straight line.

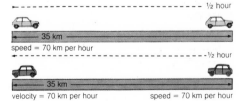

acceleration (*n*) the increase in velocity per unit time, i.e. the increase in velocity divided by the time taken to increase it. If the velocity of a motor-car increases from 20 metres per second (m/s) to 30 m/s in 5 seconds, then the acceleration = (increase in velocity) ÷ (time) = $(30 - 20)$ m/s ÷ 5 s = 10 m/s ÷ 5 s = 2 m/s² (metres per second per second) **accelerator** (*n*), **accelerate** (*v*), **accelerated, accelerating** (*adj*).

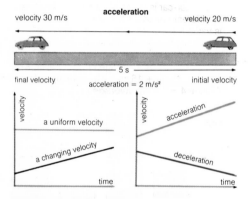

acceleration

velocity 30 m/s — velocity 20 m/s

— 5 s —

final velocity — acceleration = 2 m/s² — initial velocity

a uniform velocity

a changing velocity

acceleration

deceleration

velocity — time — velocity — time

deceleration (*n*) the rate of decrease in velocity. Deceleration is negative acceleration.
decelerate (*v*).
rest (*n*) a state of not being in motion, e.g. a motor-car is either at rest, or it is travelling (in motion).
initial (*adj*) describes a speed, a velocity, an acceleration, or any measurement that is the first to be considered, e.g. the initial length of a spring is 10 cm. The spring is stretched and its final length is 12 cm.
uniform (*adj*) describes a speed, or a velocity, or an acceleration, that does not change with time; or also any shape, or colour, or appearance, or measurement that is the same over the whole of an object, e.g. a water-pipe of uniform diameter (p.10) has the same diameter along the whole of its length.

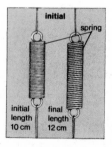

initial

spring

initial length 10 cm — final length 12 cm

quantity (*n*) (1) An indefinite amount of a material when no measurement is given, e.g. take some milk and add the same quantity of water: the quantity can be measured by volume or by weight   A *quantity* of sand. (2) The different kinds of measurements made in science. Examples of scientific quantities are: length, weight, volume, velocity, density, force.

variable (*adj*) describes a quantity which can change, or can be changed, e.g. (1) the speed of a motor-car is variable – a person can drive it at different speeds; (2) the velocity of the wind is variable for it changes by itself. **vary** (*v*).

average (*n*) an average is the sum of variable quantities divided by the number of the quantities, e.g. the average of 10 m, 16 m, 8 m, 12 m, is: $(10 m + 16 m + 8 m + 12 m) \div 4 = 46 m \div 4 = 11.5 m$. **average** (*adj*).

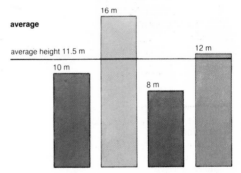

average

average height 11.5 m

standard (*n*) a measurement that is agreed by all people. Either it is clearly described, or it can be used to see if a measuring instrument is accurate, e.g. (1) the metre is the standard of length, it has an accurate description; (2) a kilogram of metal is the standard of mass (p.14) and it is used to see if a weighing instrument is accurate. **standard** (*adj*).

unit (*n*) a standard measurement, e.g. (1) a length of 12 m has a number: 12, and a standard unit: metres; (2) a volume of 0.4 m³ has a number: 0.4, and a standard unit: cubic metres.

mass (*n*) a measure of the amount of a material.
All materials possess mass. Masses are
compared by weighing which is comparing the
forces of gravitation (p.17) acting on them. It
is measured in kilograms. The mass of an object
never changes.

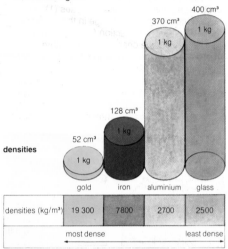

densities

densities (kg/m³)	19 300	7800	2700	2500
	gold	iron	aluminium	glass

most dense ←————————————————→ least dense

density (*n*) mass per unit volume, i.e. the mass of
a material divided by its volume, e.g. 128.2 cm³
of iron has a mass of 1 kg: density of iron =
(mass) ÷ (volume) = 1 kg ÷ 128.2 cm³ =
7.8 g/cm³ = 7800 kg/m³ (kilograms per cubic
metre). Each material has a particular density,
e.g. the density of water is 1 g/cm³, so density is
important in identification (p.93) of materials.
dense (*adj*).
relative density relative density = (density
of a material) ÷ (density of water). For iron,
density is 7800 kg/m³, for water, density is
1000 kg/m³. The relative density of iron =
7800 kg/m³ ÷ 1000 kg/m³ = 7.8. Relative
density is only a number, it has no units.
specific gravity another name for relative
density.

Volume and mass are
two variable quantities.
One depends on the other.

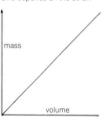

momentum

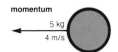

20 kg m/s momentum

20 kg m/s momentum

momentum (*n*) momentum is the product of the mass and the velocity of an object. For example, an object with mass (↑) of 5 kg moving with velocity (p.11) of 4 m/s has a momentum of 20 kg m/s. A solid, or a fluid (p.39), when in motion possesses momentum.

force (*n*) a push or pull which causes (1) acceleration, *or* (2) a change in the shape of an object, *or* (3) a reaction (p.16). A force is measured by the change in momentum produced in 1 second. A force cannot be seen, only its effects can be seen. A force can also be measured by (1) the amount it stretches a spring; (2) the acceleration it gives to a mass. Force = (mass) × (acceleration). In symbols, $F = ma$. Examples of common forces are shown:

common forces

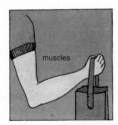

tendency (*n*) trying to continue an action, or to obtain a result or to maintain a certain direction, or situation. **tend** (*v*).

inertia (*n*) the tendency (↑) of an object to maintain its state of rest or uniform motion in a straight line. A force is needed to oppose inertia. **inertial** (*adj*).

newton (*n*) the unit of measurement for a force. A force of 1 newton gives an acceleration of 1 m/s² to a mass of 1 kg. The symbol is N.

force is measured in newtons

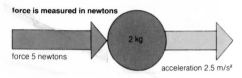

force 5 newtons

2 kg

acceleration 2.5 m/s²

action (*n*) the effect produced by a force, e.g. a hammer (p.10) hits a piece of metal, the force of the hammer acts on the metal; the action of the force is the effect: the metal changes shape. **act** (*v*).

action and reaction are equal and opposite in direction

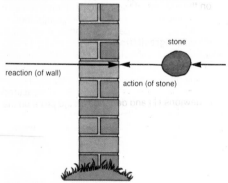

stone

reaction (of wall)

action (of stone)

reaction (*n*) the opposite effect to an action, e.g. when a stone is thrown to hit a wall, the stone has an action on the wall. The wall does not move, nor is its shape changed. Instead the wall pushes back with a reaction which is equal to the action. A reaction is equal to an action, but is in the opposite direction.

diagram (*n*) a simple drawing using mainly lines. For example, a diagram is used to explain how a device works, or how an instrument is used. Words, called labels, help the description.

attraction (*n*) a force which tries to draw two objects together, e.g. a magnet has an attraction for a compass needle. **attract** (*v*).

gravity (*n*) the attraction of the Earth for all solids, liquids, and gases, e.g. if dropped a book falls to the floor because of gravity, that is, the Earth has an attraction for the book. Gravity depends on the masses of two objects and their distance apart. The greater the mass, the greater the attraction; the greater the distance, the less the attraction. **gravitational** (*adj*).

weight (*n*) the force of gravity on a mass which attracts the mass towards the Earth. The mass of an object does not vary. It is measured in kilograms. The weight of an object is measured in newtons (↑) and depends on the place on the Earth where it is measured. A mass of 1 kg has a weight of 9.78 N at the Equator and a weight of 9.83 N at the North or South Poles. **weigh** (*v*) to measure the mass or the weight of an object.

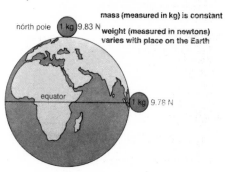

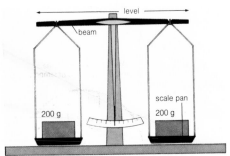

a balance (measures mass)

balance (*n*) an instrument (p.10) for measuring mass. When two equal masses are placed in the scale pans, *see diagram*, the arm of the balance is level. A balance weighs masses accurately anywhere on the Earth. **balance** (*v*).

balance (*v*) of forces in opposition, to be equal in magnitude (↓). **balance** (*n*).

spring balance an instrument for measuring weight. The weight of an object stretches the spring and the weight is read on the scale.

approximate (*adj*) describes a measurement, or a number, which is not accurate, but is near enough in magnitude (↓) to use in a calculation, e.g. the accurate speed of light is 2.997925×10^5 km/s; the approximate speed (good enough for most calculations) is 3×10^5 km/s. **approximation** (*n*), **approximate** (*v*).

significant (*adj*) describes figures which have determining value, e.g. a piece of brass has a mass of 792 g and a volume of 94 cm³; calculation of its density is 792 g ÷ 94 cm³ = 8.425532 g/cm³; 8.4 g/cm³ is accurate to 2 significant figures because there cannot be more figures in the answer than in the original measurements; the remaining figures in the answer are usually not important.

constant (*adj*) describes a quantity or a measurement which never changes in magnitude, e.g. an object falling to Earth has an acceleration which is constant at any one place; the acceleration always has the same magnitude. **constant** (*n*).

spring balance
(measures weight)

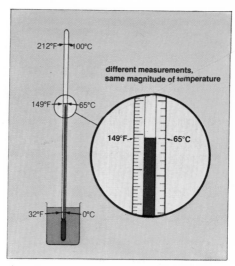

different measurements,
same magnitude of temperature

magnitude (*n*) the size of a measurement, e.g.
2 miles and 3.22 km are different measurements
of the same distance. The two measurements
have the same magnitude.

vector[1] (*n*) a quantity that has a magnitude and
a direction, e.g. velocity (p.11) and weight are
vectors. The direction of a straight line shows the
direction of the vector, its length shows the
magnitude and its origin shows the point of
action. A line is said to *represent* the vector.

scalar (*n*) a quantity that has magnitude but no
direction, e.g. mass, density.

resultant (*n*) the single force which has the same
effect as two or more forces acting on an object.

component (*n*) a force acting on an object can be
considered as two, or more, forces called
components, which produce the same effect.

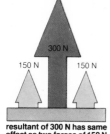

resultant of 300 N has same
effect as two forces of 150 N

pushes roller with force F

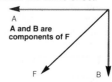

A and B are
components of F

equilibrium

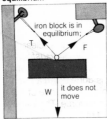

iron block is in equilibrium;

T   F

W   it does not move

triangle of forces

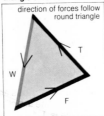

direction of forces follow round triangle

T

W

F

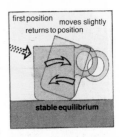

first position   moves slightly
returns to position

stable equilibrium

equilibrium (*equilibria*) (*n*) a state of balance (p.18) between opposing forces or effects. Forces which balance each other are *in equilibrium*. An object is in equilibrium if the forces acting on it are in equilibrium.

triangle of forces a triangle whose sides are parallel and proportional (p.23) to the three forces in equilibrium at a point. The directions of the forces must follow each other round the triangle.

stable (*equilibrium*) (*adj*) the state of an object if it returns to its first position after being moved slightly, e.g. a cup tilted on a table.

unstable (*equilibrium*) (*adj*) the state of an object if on being moved slightly it moves further away and the equilibrium is upset, e.g. a pencil balanced on its point.

neutral (*equilibrium*) (*adj*) the state of an object if it stays in its new position after being moved slightly, e.g. a ball on a level floor.

work (*n*) work is done when a force moves an object. The amount of work is measured by: (strength of force) × (distance moved). The distance is measured in the direction in which the force acts. The symbol for work is *w*. $w = f \times s$. The unit of work is the joule.

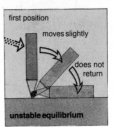

first position

moves slightly

does not return

unstable equilibrium

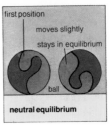

first position

moves slightly

stays in equilibrium

ball

neutral equilibrium

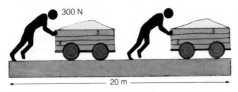

300 N

20 m

work done = 300 N × 20 m = 6000 J = 6 kJ

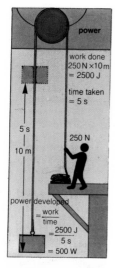

power

work done
250 N ×10 m
= 2500 J

time taken
= 5 s

5 s

250 N

10 m

power developed
= work / time
= 2500 J / 5 s
= 500 W

energy (n) the ability to do work. There are
different forms of energy: potential energy
(stored energy); kinetic energy (energy from
motion, p.11); heat energy; light energy;
electrical energy; chemical energy; nuclear
energy. One form of energy can be *transformed*
into another form. The unit of energy is the joule.

joule (n) one joule of work is done when a force
of 1 newton moves an object a distance of
1 metre in the direction of the force. The symbol
for joule is J. $1 J = 1 N \times 1 m$.

power (n) the rate (p.23) of doing work, i.e. the
amount of work done divided by the time taken to
do the work. If a force does 20 J of work in 5
seconds, then the power is: $(20 J \div 5 s) = 4 J/s$.
The symbol for power is *P*. The unit of power
is the watt.

watt (n) a power of 1 watt is produced when 1
joule of work is done in 1 second. The symbol
for watt is W. $1 W = 1 J \div 1 s$.

instantaneous (adj) happening in an instant, that
is no time seems to be taken between a cause
and its effect, e.g. when a bell is struck, no time
is taken between the striking and the ringing
of the bell.

instantaneous

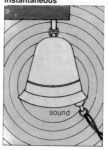

sound

simultaneous

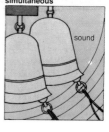

sound

simultaneous (adj) (of *two* events) happening at
the same time, e.g. two stones, held in one
hand, drop simultaneously, when released.

potential energy stored energy is potential
energy. Energy can be stored: by raising a
weight through a height; in a stretched spring.

resistance¹ (*n*) a force which opposes a change in motion or a change in shape, e.g. a fluid offers resistance to an object moving through it.

friction (*n*) the resistance to motion between two surfaces moving over each other. Friction only appears when the surfaces try to move. If the moving force is not strong enough, friction prevents motion. **frictional** (*adj*).

overcome (*v*) to be too strong for an opposition. Friction opposes the motion of an object. If a force is strong enough it overcomes friction and the object moves.

lubricant (*n*) a material put on a surface to lessen friction. Oil is used as a liquid lubricant, graphite as a solid lubricant. **lubricate** (*v*).

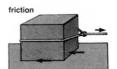
friction

friction prevents motion

lubricant

brake

brake (*n*) a device (p.10) which uses friction to stop a wheel from turning. **brake** (*v*) to use a brake to make a motor-car, cart, or engine, travel more slowly, or stop.

lever (*n*) a bar used to lift or move heavy weights. The bar turns about a point when lifting the weight. There are three types of levers: simple lever, crowbar lever and tongs lever.

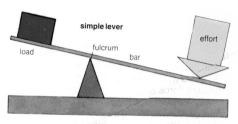

simple lever

load · fulcrum · bar · effort

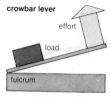

crowbar lever

effort · load · fulcrum

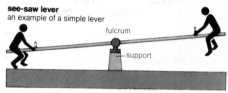

see-saw lever
an example of a simple lever

fulcrum · support

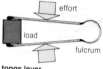

tongs lever

effort · load · fulcrum

fulcrum (*n*) the point about which a lever turns, or the support about which a lever turns.

load (*n*) (1) the weight lifted or moved by a lever, or any other machine (p.24). (2) the weight supported by a pillar; the weight on a bridge.

effort (*n*) the force applied (p.4) to a lever to lift a load; the force applied to any machine to raise a load.

exert (*v*) to bring or put into action, e.g. the earth exerts an attractive force on the moon.

input (*n*) the total amount of energy put into a device.

output (*n*) the amount of useful energy given out by a device.

ratio[1] (*n*) the relation between two numbers or two measurements, usually with the same unit, e.g. the ratio of 7 m to 1 m is written as 7:1 or 7/1.

rate (*n*) the relation between two measurements with different units, e.g. the rate of change of distance with time (i.e. speed, measured in metres/second).

proportion (*n*) the condition of a set of ratios being equal, e.g. 1 mm/7 N and 3 mm/21 N are in proportion. The lengths measured in millimetres are proportional to the forces measured in newtons. **proportional** (*adj*).

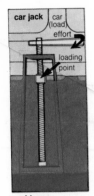

machines

machine (*n*) a device (p.10) used to help in doing
work; in it a force (the effort), applied at one
part of the machine, overcomes another force
(the load), acting at another part of the machine;
or the machine changes the direction of
application of a force. A lever is a simple machine.

mechanical advantage the number of times
the load on a machine is greater than the
effort, e.g. a load of 600 N is raised by an effort
of 150 N; the mechanical advantage of the
machine is: (600 N ÷ 150 N) = 4.

velocity ratio the ratio (↑) of the distance moved
by the effort to the distance moved by the load.

inclined plane a simple machine which is
a sloping plane surface. Friction between the
load and the inclined plane lessens the
mechanical advantage.

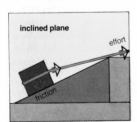

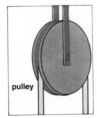

pulley (*n*) a wheel over which a rope, or chain, passes; it is used to lift weights, or to pull objects. The wheel turns on a pin fixed in a frame.

pulley-block several pulleys inside a frame.

block and tackle pulley system

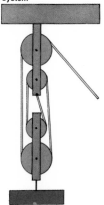

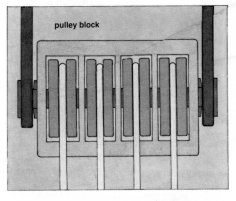

pulley block

pulley system an arrangement of pulleys to form a machine.

tension (*n*) (1) the force exerted on or by a stretched object, e.g. a rope, a chain, a spring. The rope, chain, or spring that is stretched is *under tension*. (2) the condition of a rope etc., that is being stretched. The tension in a rope, chain or spring, is the same force along the whole of its length. **tensile** (*adj*).

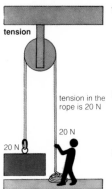

tension

tension in the rope is 20 N

20 N

20 N

revolution (*n*) (1) the motion of an object round a point outside itself, e.g. the revolution of the Earth round the Sun. (2) one complete turn of a wheel is one revolution; the circumference (p.10) is considered to revolve. **revolve** (*v*).

efficiency (*n*) the ratio (usually given as a percentage) of the energy output (p.23) to the energy input (p.23). If 200 J of work is done by the effort and 160 J of work is done on the load, then the efficiency = (160 J ÷ 200 J) × 100% = 80%. Some work is always lost because of overcoming friction in the moving parts of a machine.

perfect (*adj*) a machine which has 100% efficiency is perfect; no practical machine is perfect.

particle (*n*) a small piece of material, such as a grain (of sand), a molecule (↓) or an atom (p.103).

molecule[1] (*n*) the smallest particle (↑) of any material that has the chemical properties (p.93) of that material. Materials are considered to be made of molecules held together by attractive (p.17) forces. **molecular** (*adj*).

cohesion (*n*) the force of attraction (p.17) between molecules of the same solid, or liquid, material, e.g. the cohesion between molecules of glass. **cohesive** (*adj*), **cohere** (*v*).

adhesion (*n*) the force of attraction (p.17) between molecules of different materials, e.g. the adhesion between molecules of water and molecules of glass. The adhesion of water to glass is stronger than the cohesion of water. **adhesive** (*adj*), **adhere** (*v*).

formation of a drop of water in oil, ball-shape formed

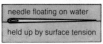

needle floating on water

held up by surface tension

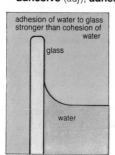

adhesion of water to glass stronger than cohesion of water

glass

water

glass

mercury

cohesion of mercury stronger than adhesion of mercury to glass

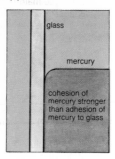

adhesion/cohesion

surface tension the tendency (p.15) of the surface of a liquid to behave as though covered with a skin. This is due to the cohesive forces on molecules at and near the surface not being balanced; such molecules are attracted more towards the centre of the liquid. The effect of surface tension is to make a drop of liquid take up the smallest space, usually in the shape of a ball.

capillarity (*n*) the rise or fall of liquids in a capillary tube; a capillary tube is a pipe with a very small diameter (p.10). This effect causes the *capillary rise* of water in plants. **capillary** (*adj*).

meniscus (*n*) the curved surface of a liquid in a pipe, or in a narrow vessel.

capillarity

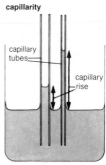

capillary tubes

capillary rise

the smaller the diameter of capillary tube the greater the capillary rise

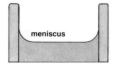

meniscus

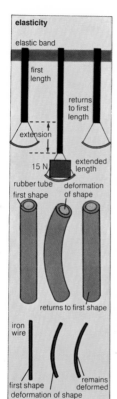

elasticity

elastic band

first length

returns to first length

extension

15 N

extended length

rubber tube
first shape

deformation of shape

returns to first shape

iron wire

first shape

remains deformed

deformation of shape

kinetic (*adj*) having to do with motion (p.11). *Kinetic energy* is the energy of motion. If an object has a mass *m* and velocity *v*, and kinetic energy *E*, then $E = \frac{1}{2}mv^2$.

Brownian motion or **movement** the continuous, irregular motion of particles in a fluid on their being hit by moving molecules of the fluid. This effect can be seen in smoke.

diffusion (*n*) the spreading of molecules of a material through a solid, a liquid or a gas. A drop of ink in water slowly diffuses throughout the whole of the water. Diffusion is caused by the motion of molecules, as seen in Brownian motion (↑).

property (*n*) a description of a material, which allows it to be recognized; *physical property* a property which does not affect the chemical nature of the material; physical properties include: size; smell; colour; density (p.14); state, i.e. solid, liquid, or gas; elasticity.

elasticity (*n*) the ability of an object, or a material, to return to its first size, or shape, after being stretched by a force, and the force then taken away. **elastic** (*adj*) both rubber and a steel wire are elastic.

extension (*n*) the increase in length when an elastic solid is stretched by a force. **extend** (*v*).

deformation (*n*) an alteration in the size or shape of a solid. Within the elastic limit (p.28) an elastic solid returns to its first size or shape when the deforming force is taken away. For a plastic solid, the deformation remains when the force no longer acts. **deform** (*v*), **deformed** (*adj*).

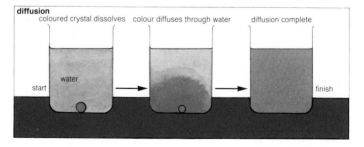

diffusion

coloured crystal dissolves   colour diffuses through water   diffusion complete

water

start   finish

stress (*n*) the force per unit area acting on a
solid. The stress produces a deformation (p.27).

strain (*n*) a change in the size or shape of a
solid produced by a stress. Strain is measured
by: (change in size) ÷ (size before stress).
If an elastic wire is 1.5 m long and a stress
causes an extension of 3 mm, then the strain is
3 mm ÷ 1500 mm = 0.002. For a particular
material, (stress) ÷ (strain) is a constant within
the elastic limit (↓).

Hooke's law strain is proportional to stress, i.e.
the extension of an elastic solid is proportional
to the force stretching the solid, e.g. if a wire has
an extension of 3 mm under a stress of
5000 N/m², then it will have an extension of
6 mm under a stress of 10 000 N/m².
3 mm/5000 N/m² and 6 mm/10 000 N/m² are *in
proportion*. There is a limit, the **elastic limit**,
beyond which the solid will no longer be elastic.

limit (*n*) a line, or a point, or a point in time,
or a magnitude, beyond which it is not possible
to go, e.g. the limit of hearing; the limit of
seeing: i.e. the greatest distance at which a
person can see an object.

restoring force (*n*) the force within a material,
which returns a stretched elastic solid to its
first length.

rigid (*adj*) describes a solid which does not change
in shape when a force acts on it, nor is it
deformed by a stress, e.g. a pillar is a rigid
solid.

plastic (*adj*) describes a solid which is deformed
(p.27) by a stress. The stress changes the
shape, or size, of the solid, and the shape
remains changed when the stress is taken away,
e.g. hot candle wax is plastic.

Hooke's law

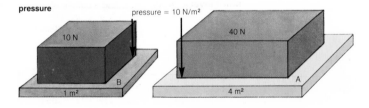

pressure (*n*) force per unit area, i.e. the force
spread over a particular area. Force is measured
in newtons and area in square metres, so
pressure is measured in newtons per square
metre. In the diagram opposite, a force of 40 N
acts on an area of 4 m² at surface A. The pressure
is 40 N/4 m² = 10 N/m². For surface B, the
pressure is 10 N/1 m² = 10 N/m². The symbol
for pressure is *p*. Pressure is *transmitted* by
solids, and fluids in vessels, from one place to
another; pressure is *applied to* a surface, a liquid,
or a gas. **press** (*v*).

pascal (*n*) the unit of pressure; the symbol is
Pa. 1 Pa = 1 N/m².

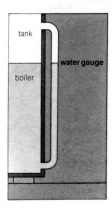

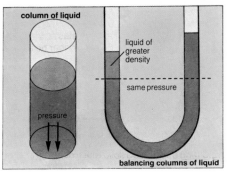

balancing columns of liquid

column of liquid liquid contained in a pipe.
The pressure on the bottom surface depends on
the height of the liquid and its density. Columns
of liquids can be balanced (p.18), and used to
find the relative density (p.14) of liquids.

water gauge a device (p.10) for measuring the
depth of water in a vessel such as a boiler.

spirit level a device containing a bubble of
air, or oil, in a liquid (usually alcohol). The
spirit level is used to test whether a surface is
horizontal.

spirit level

bubble

Pascal's law the pressure on a liquid in a
closed vessel is transmitted (*see* pressure p.29),
without becoming less, in all directions. In the
diagram, the pressure at A (40 N/cm²) causes a
pressure of 40 N/cm² on B.

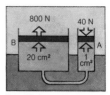

Pascal's law

hydraulic press a machine using hydraulic
pressure. The effort (200 N) produces a pressure
of 40 N/cm² on the large piston at B. Force =
(pressure) × (area). The force produced at B =
40 N/cm² × 500 cm² = 20 000 N. The mechanical
advantage (p.24) is 20 000 N ÷ 200 N = 100. A
hydraulic jack and *hydraulic brakes* act in the
same way as a hydraulic press.

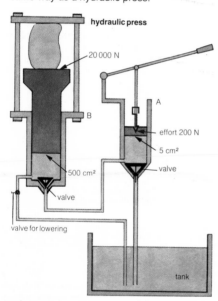

hydraulic press

20 000 N

A

B

effort 200 N

5 cm²

valve

500 cm²

valve

valve for lowering

tank

atmospheric pressure the pressure (p.29)
caused by the air in the Earth's atmosphere on
the surface of the Earth. The pressure varies
continuously from day to day. Normal pressure
(i.e. not high, not low) is taken as 101 325 Pa.
The approximate value is 101 kPa. This will
support a column of mercury 760 mm high.

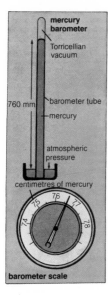

mercury barometer

Torricellian vacuum

760 mm — barometer tube
— mercury

atmospheric pressure

centimetres of mercury

barometer scale

barometer (*n*) an instrument for measuring atmospheric pressure. A mercury barometer, *see diagram*, uses mercury, a liquid metal, in a tube closed at the top, with the lower end in a vessel containing mercury. **barometric** (*adj*).

Torricellian vacuum the space above the column of mercury in a barometer. There is nothing in the vacuum.

vacuum (*n*) a space with nothing in it; this is an *absolute* vacuum. A space from which as much air (or other gas) as possible has been taken; this is a *high* vacuum. Taking less air produces a *low* vacuum, also called a *partial* vacuum.

aneroid barometer a thin circular metal box, from which air has been taken, (leaving a partial vacuum) measures atmospheric pressure. An increase in atmospheric pressure causes the box to bend inwards; this movement is passed on by levers so that a needle turns on the scale, *see diagram*.

aneroid barometer

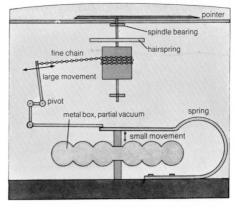

pointer

spindle bearing

fine chain — hairspring

large movement

pivot

metal box, partial vacuum — spring

small movement

altimeter (*n*) an aneroid barometer suitably altered to measure height, by measuring the change in atmospheric pressure with height. A mercury barometer falls approximately 1 mm for a rise in height of 12 m.

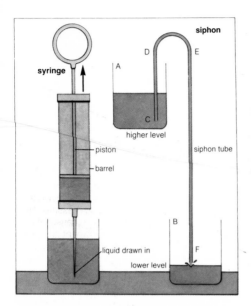

suction (*n*) the drawing in of a fluid to produce a
   partial vacuum which is filled by atmospheric
   pressure. **suck** (*v*).

syringe (*n*) a device (p.10) for sucking in liquid
   and then pushing it out under pressure. A
   syringe has a barrel in which a piston moves up
   and down. When the piston (*see diagram*) is
   raised, atmospheric pressure pushes liquid in.
   When the piston is lowered, the liquid is pushed
   out.

siphon (*n*) a device for taking liquid out of a
   vessel and passing it to a lower level. A curved
   tube, *see diagram*, full of liquid, is used. The
   side EF must be longer than the side CD for
   the tube to siphon water out of vessel A and
   pass it to vessel B.

manometer (*n*) an instrument for measuring small
   pressures of gases. A U-tube contains liquid; the
   pressure of the gas pushes the liquid (usually
   water or mercury) round the tube.

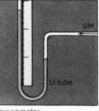

manometer

valve[1] (*n*) a device (p.10) which: (1) controls the flow of a liquid or a gas through a pipe, or (2) allows liquid or gas to flow through a pipe in one direction only.

pump (*n*) a machine for: (1) taking liquids or gases from one place and passing them to another place; (2) forcing gases into smaller spaces. A pump uses valves and pistons.

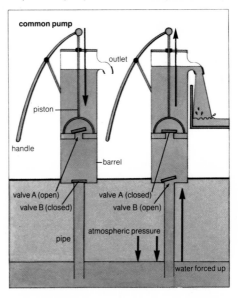

common pump

outlet

piston

handle

barrel

valve A (open)
valve B (closed)

valve A (closed)
valve B (open)

atmospheric pressure

pipe

water forced up

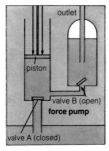

outlet

piston

valve B (open)

force pump

valve A (closed)

common pump (*n*) a machine used to raise water or other liquids. A piston moves up and down in the pump barrel, *see diagram*, the pump draws liquid into its barrel and pushes the liquid out through an outlet. This pump cannot raise water more than 10 m.

lift pump another name for common pump.

force pump a pump that draws liquid into its barrel (by atmospheric pressure) but forces the liquid out under pressure to a height greater than 10 m depending on the force acting on the piston.

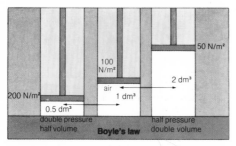

Boyle's law the law that using a particular
mass of gas, at a fixed temperature:
(pressure of gas) × (volume of gas) is constant.
In the diagram: (200 N/cm² × 0.5 dm³) =
(100 N/m² × 1 dm³) = (50 N/dm³ × 2 dm³).
Boyle's law in symbols is pV = K (a constant).

Bernouilli's principle when liquid flows
through a pipe of varying diameter, the amount
of energy per kilogram of liquid does not change.
When the velocity of flow is greatest, the
pressure is least. Liquid flows faster through a
narrow part of the tube, and slower through a
wider part.

Venturi tube a tube which shows Bernouilli's
principle. Two wide diameter tubes are joined by
a narrow tube. Mercury manometers (p.32)
show the pressure is low in the narrow tube and
higher in the wide tubes.

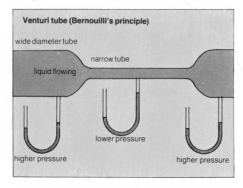

streamline flow

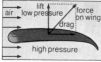

turbulent flow

aerofoil (aeroplane wing)

streamline (*adj*) describes the flow of a liquid, or gas, which has no sudden change of direction.

streamlined (*adj*) describes the shape of a solid such that any liquid or gas passing it will have streamline flow.

turbulent (*adj*) describes the flow of a liquid, or gas, in which the flow changes rapidly in direction and magnitude. Turbulent is the opposite of streamline.

aerofoil (*n*) a surface, e.g. of an aeroplane which helps the aeroplane to remain in the air, or to change direction, e.g. wings are aerofoils.

suspend (*v*) (1) to hang from a support, e.g. to suspend a load by a rope from a pulley; (2) to be held or to remain in a position, e.g. dust suspended in the air. **suspended** (*adj*).

immerse (*v*) to put a solid object partially, or wholly, under the surface of a liquid, usually water, *see diagram*.

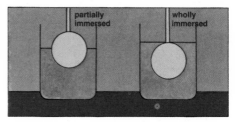

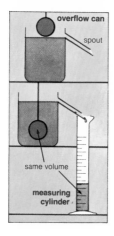

submerge (*v*) to put a solid object completely into water, usually at a depth.

displacement (*n*) (1) the volume (or weight) of fluid pushed out of place by an object; (2) the difference between the first position of an object and any later position, *see diagram*. **displaced** (*adj*), **displace** (*v*).

overflow can a metal can with a spout. The can is filled with water and an object submerged in the water. The object displaces water through the spout. The volume of displaced water equals the volume of the object.

measuring cylinder a tall glass vessel which measures, in cubic centimetres, the volume of a liquid.

upthrust (*n*) a force acting in an upward direction.

Archimedes' principle when an object is wholly, or partially, immersed (p.35) in a liquid, there is an apparent loss in weight. The loss in weight is caused by an upthrust on the object. The upthrust is equal to the weight of displaced (p.35) liquid. Archimedes' principle can be used for gases as well as liquids.

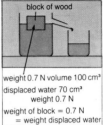

law of flotation

block of wood

weight 0.7 N volume 100 cm³
displaced water 70 cm³
weight 0.7 N
weight of block = 0.7 N
= weight displaced water

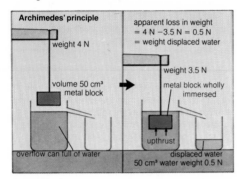

Archimedes' principle

weight 4 N

volume 50 cm³
metal block

overflow can full of water

apparent loss in weight
= 4 N −3.5 N = 0.5 N
= weight displaced water

weight 3.5 N

metal block wholly
immersed

upthrust

displaced water
50 cm³ water weight 0.5 N

flotation (*n*) a floating object displaces its own weight of the fluid in which it is floating. This is the law of flotation; it is a special case of Archimedes' principle. **float** (*v*).

buoyancy (*n*) (1) the upthrust of a fluid on an object which is immersed (p.35) in it; (2) the tendency (p.15) of an object to float in a fluid. Buoyancy decides whether a solid object floats or sinks. **buoyant** (*adj*) (1) able to float; (2) able to keep an object floating.

balloon (*n*) a bag-like object filled with a gas of low density (p.14). The bag is made of a soft, but strong, material. The balloon displaces air; if the upthrust of the displaced air is greater than the weight of the balloon, then the balloon rises in the air.

hydrometer (*n*) a floating instrument (p.10) used for measuring the density (p.14) of a liquid. It has a hollow glass body with a long stem, and a weight at the bottom to make it float upright. The stem shows a scale of densities; the higher densities are at the bottom of the scale.

balloon

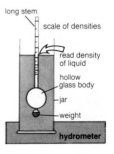

long stem

scale of densities

read density
of liquid

hollow
glass body

jar

weight

hydrometer

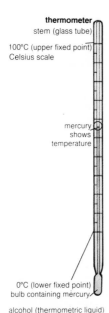

thermometer
stem (glass tube)
100°C (upper fixed point) Celsius scale

mercury shows temperature

0°C (lower fixed point) bulb containing mercury

alcohol (thermometric liquid)
space for expansion
bulb

index
maximum temperature recorded here

minimum temperature recorded here

column of mercury

maximum and minimum thermometer

temperature (*n*) a measure, using a scale, of how hot, or how cold, an object, an organism (p.147), or the atmosphere is.

heat (*n*) a form of energy which materials possess from the kinetic energy of their molecules (p.26); heat is measured in joules (p.21). The physical effects of heat are: (1) to change the temperature of a material; (2) to change the state (p.39) of a material; (3) to cause expansion (p.38). Heat is transferred (p.45) from a material at a higher temperature to one at a lower temperature. A change in heat content of an object is measured by its heat capacity (p.42) multiplied by the change in temperature. **heat** (*v*).

thermometer (*n*) any instrument for measuring temperature; it uses a physical property of a material that changes regularly with a change in temperature. The commonest kind uses the expansion of mercury in a glass tube, *see diagram*; other thermometric liquids, such as alcohol, can be used instead of mercury. **thermometric** (*adj*).

fixed point a standard temperature. The upper fixed point on thermometer scales is the boiling point (p.41) of pure water; the lower fixed point is the temperature of melting ice (p.20).

Celsius scale a temperature scale with 100 degrees between the lower fixed point (0°C) and the upper fixed point (100°C). The symbol is: °C.

kelvin (*n*) the S.I. unit of temperature (↑); the symbol is: K. 1 K = 1°C. The lower fixed point is 273 K and the upper fixed point is 373 K.

maximum and minimum thermometer a thermometer which measures the highest (maximum) and the lowest (minimum) temperatures recorded during a particular length of time, usually one day, *see diagram*. The thermometric (↑) liquid is alcohol and it pushes a column of mercury round a U-shaped thermometer. The mercury has two metal indices to record (p.41) the maximum or the minimum temperatures. The indices are reset by a magnet.

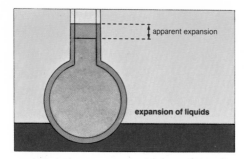

apparent expansion

expansion of liquids

expand (*v*) to increase in length, area, or volume of a solid, or in volume of a fluid. The increase is caused by a rise in temperature, and for gases is also caused by a decrease in pressure (Boyle's law p.34). **expansion** (*n*).

apparent expansion the measured expansion of a fluid in a vessel. On heating, both the vessel and the fluid expand, so the true expansion of the fluid is its apparent expansion added to the expansion of the vessel.

contract (*v*) to decrease in length, area, or volume of a solid, or in volume of a fluid. The decrease is caused by a fall in temperature and for gases is also caused by an increase in pressure (Boyle's law p.34), e.g. a metal rod contracts on cooling. **contraction** (*n*).

coefficient (*n*) a constant ratio which measures a particular property for a change in quantity of a material, e.g. the coefficient of linear expansion is the increase in length per metre of a solid for a rise in temperature of 1°C.

bimetallic (*adj*) made of two different metals, e.g. iron and brass. A bimetallic strip bends when heated, as the two metals have different coefficients of expansion.

thermostat (*n*) a device for keeping a fluid, or an object, at a constant temperature. The bimetallic strip of brass and iron, *see diagram,* bends when hot and straightens when cold. When cold it makes an electric contact to start a heating coil. When sufficiently hot, the contact is broken.

expansion of metals

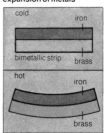

cold
iron
bimetallic strip
brass
hot
iron
brass

simple thermostat

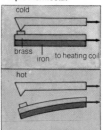

cold
brass
iron
to heating coil
hot

state of matter all materials are solids, liquids, or gases. These are the three states of matter.

solid (*n*) a solid possesses molecules (p.26) held together by strong forces, so solids have a definite shape and a definite volume. **solid** (*adj*), **solidify** (*v*).

liquid (*n*) a liquid possesses molecules held together by weaker forces than those in solids, so a liquid has a definite volume but no shape. **liquid** (*adj*), **liquefy** (*v*).

gas (*n*) a gas possesses molecules which are free to move about with no forces holding them together, so a gas has no definite volume and no shape. **gaseous** (*adj*).

fluid (*n*) any material that flows, i.e. a liquid or a gas. **fluid** (*adj*).

Charles' law the volume of a fixed mass of gas at a constant pressure is proportional to its temperature (measured in kelvin). In symbols: $V \propto T$.

gas law Boyle's law and Charles' law are made into one law. For a fixed mass of gas: (pressure) × (volume) is proportional to (temperature); the temperature is measured in kelvin. In symbols: $pV \propto T$.

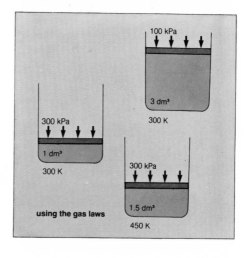

100 kPa

3 dm³

300 K

300 kPa

1 dm³

300 K

300 kPa

1.5 dm³

450 K

using the gas laws

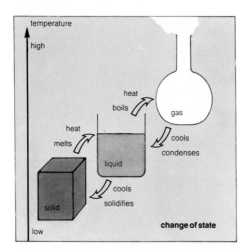

change of state

change of state the change of a material from
one state to another, e.g. from a solid to a
liquid, or a liquid to a gas. Such changes are
mainly caused by heating or cooling.

melt (*v*) to change, when heated at a constant
temperature, from a solid to a liquid, e.g. ice
melts to form water. Compare dissolve (p.89).
melted (*adj*), **molten** (*adj*).

solidify (*v*) to change, when cooled, from a liquid
to a solid, e.g. liquid wax, on cooling, becomes
solid. Compare crystallization (p.110) and
sublimation (p.92).

melting point the temperature at which a solid
melts. Each solid has its own melting point,
e.g. the melting point of ice is 0°C, of copper is
1083°C. It is a physical property useful for
identification (p.93) of materials. Abbreviation:
m.p.

freezing point the temperature at which a liquid
solidifies. For a pure material, it is the same
temperature as the melting point.

boil (*v*) to change, when heated at a constant
temperature, from a liquid to a vapour (↓), e.g.
water boils at 100°C to form steam; alcohol
boils at 78°C to form alcohol vapour.

**determination
of melting point**

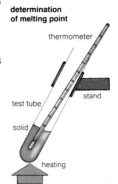

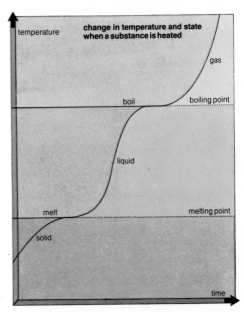

change in temperature and state
when a substance is heated

temperature

gas

boil — boiling point

liquid

melt — melting point

solid

time

boiling point the temperature at which a liquid
boils. Each liquid material has its own boiling
point, e.g. the boiling point of water is 100°C; of
alcohol is 78°C at normal atmospheric pressure.
It is a physical property useful for identification
(p.93) of materials. Abbreviation: b.p.

vapour (*n*) a vapour is a gas which can be
liquefied by increasing the pressure without
changing the temperature. It is the gaseous form
of a material which is a solid or a liquid at
room temperatures, e.g. petrol vapour; water
vapour. **vaporize** (*v*).

record (*v*) (1) to write a reading taken from an
instrument, or an observation (p.93), e.g. to
record room temperature; to record the hours of
sunshine; (2) for an instrument, to keep showing
the same reading, e.g. the index of a
thermometer records the maximum (p.37)
temperature during a day. The record, in both
cases, can always be seen. **record** (*n*).

evaporation (*n*) the change of a liquid to a vapour
at a temperature below, or at, its boiling point,
e.g. (a) the evaporation of rain water without the
water boiling; (b) when a salt solution is boiled
(p.40), the liquid evaporates leaving the salt.
evaporate (*v*).

condensation (*n*) the change of a vapour, or gas,
to a liquid when cooled, e.g. the condensation
of steam on a cold surface. Compare liquefaction,
the changing of a vapour to a liquid by pressure
alone.

calorie[1] (*n*) a former unit of heat. One calorie is the
quantity of heat needed to raise the temperature
of 1 g of water by 1°C. 1 calorie = 4.18 J.

calorimeter (*n*) any apparatus (p.88) used for
measuring quantities of heat, usually by finding
the rise in temperature of a known mass of
water. The simplest calorimeter is a metal vessel
used with a thermometer. **calorimetric** (*adj*),
calorimetry (*n*).

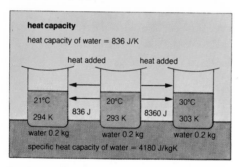

heat capacity

heat capacity of water = 836 J/K

heat added · heat added

21°C / 294 K / water 0.2 kg — 836 J — 20°C / 293 K / water 0.2 kg — 8360 J — 30°C / 303 K / water 0.2 kg

specific heat capacity of water = 4180 J/kgK

heat capacity the number of joules (p.21) needed
to raise the temperature of an object by 1 K (1°C),
e.g. the heat capacity of a copper vessel is
40 J/K; 240 J are needed to raise its temperature
by 6 K.

specific heat capacity the number of joules
needed to raise the temperature of 1 kg of a
substance by 1 K, e.g. the specific heat capacity
of copper is 400 J per kg per kelvin. If a copper
calorimeter has a mass of 0.1 kg, its heat capacity
is 40 J/K.

latent heat the heat needed to change the state of matter (p.39) of a material. While the latent heat is given to the material, its temperature remains constant, *see diagram*.

specific latent heat the number of joules needed to cause a change of state, without change of temperature, for 1 kg of a substance; it is a physical property of a substance. Each substance can have two specific latent heats: (1) of fusion (or melting); (2) of vaporization. These two quantities are constant for a particular substance, e.g. the specific latent heat of fusion of water is 336 kJ/kg and of vaporization of water is 2260 kJ/kg.

refrigeration (*n*) the use of energy to take heat away from an object. In a **refrigerator**, vapour is compressed by a pump; liquid is formed when the vapour is cooled and passed to an evaporator. The liquid takes heat from the refrigerator to supply the latent heat of vaporization, thus cooling the food in the refrigerator, *see diagram*.

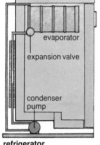

evaporator

expansion valve

condenser
pump

refrigerator

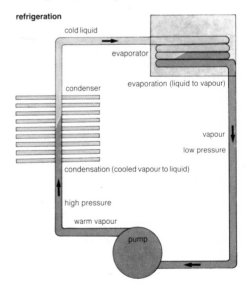

refrigeration

cold liquid

evaporator

evaporation (liquid to vapour)

condenser

vapour

low pressure

condensation (cooled vapour to liquid)

high pressure

warm vapour

pump

engine (*n*) a device which uses the properties of a working fluid to transform heat and other forms of energy into mechanical energy, e.g. a steam engine (↓) uses heat for energy and uses water as the working fluid and supplies mechanical energy.

cylinder (*n*) part of an engine in which a vapour expands, *see diagram*. Holes in the cylinder, closed by valves, allow the vapour to enter and leave after expansion.

piston (*n*) a solid circular piece of metal which can move up and down in a cylinder. Expansion of a vapour forces the piston down, thus producing mechanical energy.

steam engine heat energy boils water and produces steam under pressure; the steam expands in a cylinder, forcing the piston down. The expanded steam is led away and the piston returns. This movement is repeated. The up and down motion of the piston is changed to a circular motion by a crank, *see diagram opposite.*

reciprocating engine an engine which uses a cylinder, piston and crank to produce circular motion of a wheel, e.g. a steam engine. The motion of the wheel returns the piston.

turbine (*n*) an engine in which a shaft, *see diagram opposite*, is turned by the force of a stream of fluid on blades fixed to the shaft. In a steam turbine, steam from pipes is directed onto the blades.

internal combustion engine a mixture of petrol vapour and air enters the cylinder of the engine and is exploded by a spark. Heat from the explosion makes the gases expand and force the piston down. A four-stroke engine has (1) down-stroke; petrol mixture sucked in; (2) up-stroke; mixture compressed; (3) down-stroke; mixture exploded; (4) up-stroke; hot gases pushed out.

conservation (*n*) keeping a quantity constant, i.e. unchanging; preventing loss or waste. *Conservation of mass:* materials cannot be made or destroyed, they can only be transformed into other materials. *Conservation of energy:* energy cannot be made or destroyed, it can only be transformed. **conserve** (*v*).

internal combustion engine

1st stroke
petrol vapour sucked in

2nd stroke
vapour compressed

3rd stroke
spark explosion
working stroke

4th stroke
hot gases pushed out

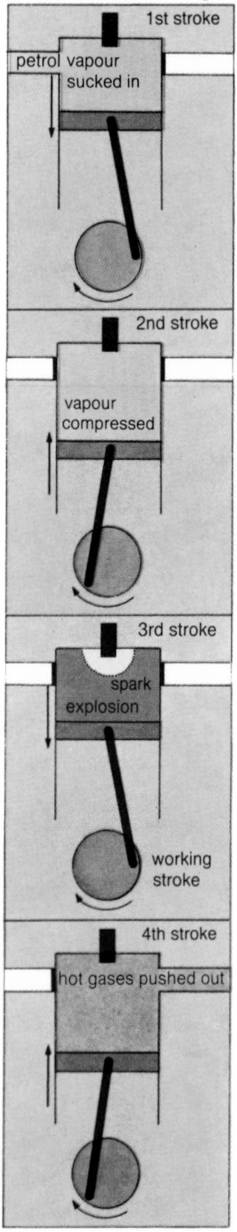

steam engine

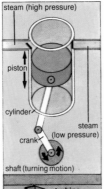

steam (high pressure)

piston

cylinder

crank

steam
(low pressure)

shaft (turning motion)

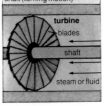

turbine

blades

shaft

steam or fluid

transfer (v) to move an object, or anything else, from one place to another by any means, e.g. to transfer heat from a hot liquid to a cold vessel. **transfer** (n).

conduction (n) the passing of heat through a solid. There is hardly any conduction of heat in fluids. Also the passing of an electric current (p.74) through a substance. **conduct** (v), **conductor** (n).

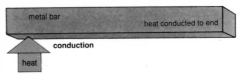

metal bar

heat conducted to end

conduction

heat

non-conductor (n) a substance that does not conduct heat or an electric current (p.74). Most non-metals are non-conductors, while metals are good conductors.

convection (n) the transfer of heat in a fluid by the rising of hot fluid and the sinking of cold fluid to take the place of the hot fluid. A **convection current** is formed by the movement of the fluid.

ventilation (n) the use of convection currents (↑) to supply a building with fresh air or with cool air. **ventilator** (n), **ventilate** (v).

radiation[1] (n) the transfer of heat from a hot object through space to a cold object, e.g. the radiation of heat from the sun to the Earth. Such heat is called *radiant heat*; it does not need any material for the transfer of heat. Compare conduction and convection, which need a material for the transfer of heat.

emission (n) the sending out of radiant heat (↑). Also the giving out of light, sound, radio waves, other kinds of radiation (p.83), and electrons (p.106), e.g. the emission of heat from the sun; the emission of light from a lamp. **emit** (v), **emitter** (n), **emissive** (adj).

absorption (n) the taking in of energy or fluids, e.g. (a) the absorption of radiant heat (↑) by a surface; (b) the absorption of carbon dioxide gas by sodium hydroxide solution. **absorb** (v), **absorptive** (adj), **absorbent** (n).

convection current

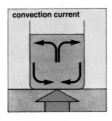

radiation

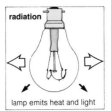

lamp emits heat and light

thermal (*adj*) of heat, e.g. (a) the thermal efficiency of an engine is the ratio of the work done by the engine to the heat energy supplied by the fuel (p.129); (b) a thermal current is a stream of air rising rapidly from a hot surface on the earth.

sea breeze a wind produced by a convection current (p.45) during the daytime. When the sun shines, land heats up more quickly than the sea. So hot air rises over the land and cooler air from the sea takes its place, thus forming a sea breeze.

land breeze a wind produced by a convection current during the evening. The land cools more quickly than the sea when the sun goes down. So hot air rises over the sea and cool air from the land takes its place; this forms a land breeze.

trade wind a wind that blows in a fixed direction for a particular season or for the whole year.

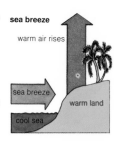

sea breeze

warm air rises

sea breeze

warm land

cool sea

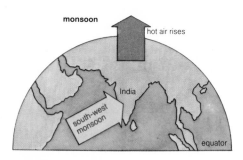

monsoon

hot air rises

India

south-west monsoon

equator

monsoon (*n*) a regular wind blowing in one direction for one part of the year and in the opposite direction for the remaining part of the year. In Asia, during summer, the land around the Arabian Sea and the Indian Ocean is very hot, and the sea is cool, so a convection current, similar to a sea breeze, is formed; this is the south-west monsoon. In winter the sea is warmer than the land, so the wind blows in the opposite direction, from the north-east; this is the north-east monsoon.

cirrus

cumulus

nimbus

stratus

whirlwind (*n*) air moving in a circular motion round a centre of low pressure. The low pressure is caused by a convection current (p.45) from a small area of very hot land, as found in deserts. A whirlwind is small in size and may suck up sand and dust as it moves over the surface of the Earth.

cyclone (*n*) (1) a large area of low pressure over part of the Earth. The pressure is lowest at the centre. Winds circle round and into the area of low pressure; rain usually falls; (2) a violent wind in the tropics caused by a smaller area of low pressure. **cyclonic** (*adj*).

wind scale a scale describing the speed of wind. The highest speed is 12, the lowest is 1.

Beaufort scale another name for wind scale.

hurricane (*n*) a wind with a speed of more than 120 km/hr; the strongest wind in the Beaufort scale (scale 12).

typhoon (*n*) a wind with a speed of more than 120 km/hr. It is the same as a hurricane (↑). The name typhoon is used in the Pacific Ocean, and the name hurricane is used in the Atlantic Ocean.

cloud (*n*) a large number of very small drops of water formed by water vapour condensing when warm air moves up to a higher level where the temperature is lower. The bottom of a cloud is at a height where the air temperature is that of the dew point (p.50).

cirrus (*n*) white clouds, shaped like feathers, at a height of 7.6 to 10 km, which are sometimes separate, and sometimes arranged regularly in groups. They are formed from ice crystals.

cumulus (*n*) thick clouds, looking like cotton wool, with a definite shape, and generally with a flat base. They are formed at a height of 3 to 5 km by warm air rising, becoming cooler and water vapour condensing.

nimbus (*n*) thick clouds, usually grey, with no definite shape, forming at a height of about 2 km. Rain or snow usually comes from nimbus clouds.

stratus (*n*) low clouds, formed in layers, usually covering a large area of sky. They are formed in calm weather; if they become thicker, rain may fall.

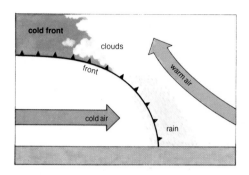

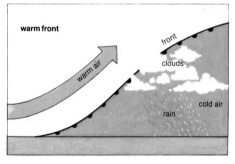

cold front a cold front is formed when cold air moves into an area of warm air. The cold air forces the warm air to rise and cumulus clouds are formed.

warm front a warm front is formed when warm air moves into an area of cold air over which it rises. Cloud is formed in layers, slowly changing from white to grey, and becoming thicker. Rain usually falls.

ridge of pressure a long, narrow area of high atmospheric pressure which comes from a greater area of high pressure.

anti-cyclone a large area of high pressure over part of the Earth; the pressure is highest at the centre. Winds circle round and out of the area of high pressure. The weather is usually clear and bright.

humidity (*n*) a measure of the amount of water vapour in the atmosphere. Usually the *relative humidity* is measured as a percentage; 100% indicates atmospheric air saturated (p.89) with water vapour and 0% indicates completely dry air. **humid** (*adj*), **humidify** (*v*).

hygrometer (*n*) an instrument for measuring the humidity of the atmosphere, e.g. the hair hygrometer which uses the expansion and contraction of hair with changes in the humidity.

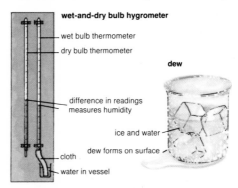

wet-and-dry bulb hygrometer

— wet bulb thermometer

— dry bulb thermometer

dew

difference in readings measures humidity

ice and water —

dew forms on surface ↖

— cloth

— water in vessel

wet-and-dry bulb hygrometer this instrument has two thermometers. One thermometer has a wet cloth round its bulb, and records a low temperature because of evaporation of the water. The dry bulb records atmospheric temperature. The difference between the readings of the two thermometers is a measure of the humidity.

precipitation[1] (*n*) the forming of drops of liquid from a vapour when the temperature has fallen below a certain value and the drops are large enough to fall. The vapour is mixed with a gas, e.g. water vapour in atmospheric air where precipitation happens when the temperature falls below the dew point (p.50), and water falls as rain or snow. **precipitate** (*v*).

dew (*n*) drops of water formed on solid surfaces by condensation (p.42) of water vapour from the air, e.g. dew is formed on grass during the night.

dew point the temperature at which the air becomes saturated (p.89) with water vapour and
condensation takes place, forming dew. Above
the dew point, atmospheric air is unsaturated.

mist (*n*) rapid cooling of the air causes very small
drops of water to form by condensation (p.42);
this is a mist. A mist is a cloud formed at
ground level. If the water condenses on dust
in the air, a fog is formed.

fog (*n*) – see mist (↑).

rain (*n*) water drops formed by precipitation,
falling from clouds.

rain-gauge (*n*) an instrument for measuring the
amount of rain which falls in a certain time, e.g.
in 24 hours. The rain is caught by a funnel and
passed into a bottle, *see diagram*. The water is
measured in a measuring cylinder, which gives a
reading of the rainfall.

rainfall (*n*) the depth, in cm, of rain water in a
given time if it did not flow away.

snow (*n*) small crystals (p.110) of ice falling from
clouds. Water drops form ice crystals if the air
temperature is below 0°C.

ice (*n*) formed from water at 0°C, when water
becomes solid. Ice forms on rivers and lakes,
less often on the sea.

rain gauge

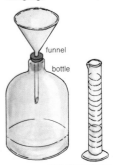

funnel

bottle

measuring cylinder

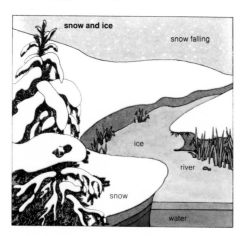

snow and ice

snow falling

ice

river

snow

water

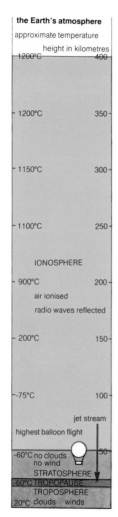

the Earth's atmosphere

approximate temperature

height in kilometres

1200°C	400
1200°C	350
1150°C	300
1100°C	250
IONOSPHERE	
900°C	200
air ionised	
radio waves reflected	
200°C	150
-75°C	100
	jet stream
highest balloon flight	
-60°C no clouds no wind	50
STRATOSPHERE	
-60°C TROPOPAUSE	
TROPOSPHERE	
20°C clouds winds	

atmosphere (*n*) the gases around the Earth, or any other heavenly body. **atmospheric** (*adj*).

troposphere (*n*) the layer of the atmosphere nearest the Earth; it varies in height from 10 km at the Poles to 18 km at the equator. The temperature falls with height above the Earth. In this layer, clouds are formed, and there are many convectional air currents from unequal heating of the Earth's surface.

tropopause (*n*) a thin layer of the atmosphere, about 5 km thick, between the troposphere and the stratosphere. The temperature no longer falls with height; fast air currents, called **jet streams**, are formed in this layer.

stratosphere (*n*) an upper layer of the atmosphere, about 25 km thick, above the tropopause. The air temperature does not increase with height; it is about –80°C over the equator and about –40°C over the Poles. No clouds or air convection currents are formed in the layer.

ionosphere (*n*) an upper layer of the atmosphere, about 350 km thick, above the stratosphere, starting at an average height of 50 km above the Earth. In it the air is ionized (p.101) by ultra-violet light (p.83) from the sun, and it reflects radio waves. The ionosphere is separated from the stratosphere by an **ozone layer**.

space (*n*) above the ionosphere is space. It contains no gases. The sun and its planets occupy a part of space and stars, galaxies, comets and all the stellar systems are found here. Only electromagnetic waves travel through space, e.g. light waves, radio waves etc.

climate (*n*) the conditions of an area on the Earth's surface such as the variation in temperature, rainfall, and humidity. There are four important types of climate: tropical; sub-tropical; temperate; polar. **climatic** (*adj*).

forecast (*v*) to say what events will happen after having seen past events, e.g. to forecast the weather, after taking measurements of temperature, humidity and wind changes. Forecasting is not accurate, but it is not expected to be very far wrong.

source (*n*) the place from which light, sound, other forms of energy, or materials come, e.g. (a) a lamp is a source of light; (b) a fire is a source of heat; (c) some minerals are the source of metals.

medium (*n*) the means by which a wave motion (p.65) travels. A medium can be material, e.g. a gas, a liquid, or a solid, or it can be non-material, i.e. a vacuum. Light does not need a material medium; it travels through a vacuum. Sound needs a material medium.

sources of light

propagation (*n*) the sending of a wave motion by a medium (↑), e.g. (a) the propagation of light by a vacuum; (b) the propagation of sound waves by air. **propagate** (*v*).

rectilinear (*adj*) formed of straight lines, travelling in a straight line, e.g. the rectilinear propagation of light is the sending of light waves in a straight line by any medium.

source of sound

transparent (*adj*) describes any solid or liquid medium through which light can travel to form an image (p.56), e.g. the glass in a window is transparent.

translucent (*adj*) describes any solid or liquid medium through which light can travel, but no clear image can be formed, e.g. waxed paper is translucent but not transparent.

transparent

opaque

translucent

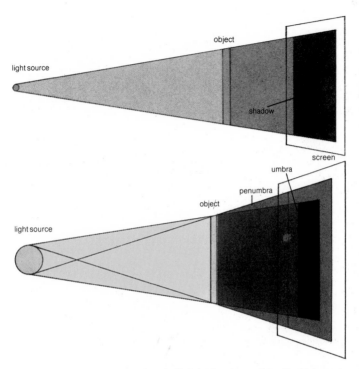

opaque (*adj*) describes any solid or liquid through which light cannot travel.

shadow (*n*) the dark area formed by an object which stops light. A shadow is formed because light travels in straight lines. The shadow has the same shape as the object when the shadow is seen on a screen (p.56). If the light source is very small, the shadow will be sharp.

umbra (*n*) the area in a shadow from which light is completely cut off.

penumbra (*n*) a lighter area between the umbra and the edge of a shadow; some light reaches it because the source is not small. A very small light source forms an umbra only; other sources form an umbra and a penumbra.

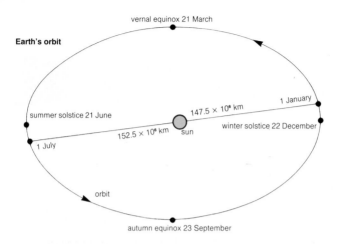

Earth (*n*) the third planet (↓) at an average distance of 150 000 000 km from the sun. It is almost spherical in shape, being slightly flattened at the Poles. The equatorial radius is 6378.388 km; the mass is 5.976×10^{24} kg.

orbit (*n*) the path followed by one object moving round another object, e.g. the orbit of the Earth round the sun. Gravitational (p.17) attraction between the Earth and the sun keeps the Earth in its orbit; the orbit takes 365¼ days to complete.

moon (*n*) a body in orbit round the Earth, held in orbit by gravitational attraction between it and the Earth. The distance from the Earth to the moon is 384 400 km; the mass is 7.35×10^{22} kg, and it takes 28 days to complete one orbit. As the Earth is itself in orbit, the path of the moon is as shown in the diagram. The moon has neither water nor atmosphere, is cold, and reflects (p.56) light from the sun.

sun (*n*) the source of light and heat for the Earth. It is spherical, its diameter is about 1 392 000 km, and its mass about 2×10^{30} kg. The surface is at a temperature of about 6000°C.

eclipse of the moon

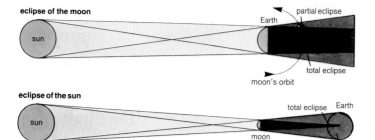

eclipse of the sun

eclipse (*n*) the darkening of a heavenly body when it moves into the shadow of another heavenly body. An eclipse of the moon is seen when the shadow of the Earth falls on the moon; an eclipse of the sun is seen when the shadow of the moon falls on the Earth.

total eclipse the condition when a heavenly body is completely in the umbra (p.53) of another heavenly body so that no light falls on it at all.

partial eclipse only part of a heavenly body is in the umbra of another heavenly body, so part of it receives light.

phase (*n*) the state, position, or condition of an object or a quantity which passes through a number of events in order, and then repeats the same events again and again, e.g. the phases of the moon are (1) new moon (2) first quarter (3) full moon (4) third quarter (1) new moon, and so on. *See diagram.*

one lunar month

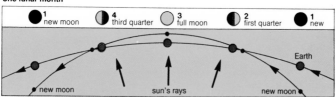

planet (*n*) a body in orbit round the sun. The Earth is a planet. Planets are not luminous (↓). (See back endpapers.) **planetary** (*adj*).

luminous (*adj*) giving out light.

angle (*n*) a measurement of a change in direction, measured in degrees or radians. There are 360° or 2π radians in a complete circle. **angular** (*adj*).

reflect (*v*) to change the direction of a line or path by means of a surface, e.g. the surface of water reflects light from the sun. **reflection** (*n*).

incident (*adj*) meeting, hitting, or falling on a surface, e.g. an incident ray of light. **incidence** (*n*).

mirror (*n*) a very smooth surface which reflects light, e.g. a metal plate or a glass plate with silver on the back.

plane mirror a flat mirror, the common kind of mirror.

ray (*n*) (1) any one line of a number of lines starting from the same point; (2) a line representing the direction of light; (3) a line of particles, in motion one after the other, going in a particular direction.

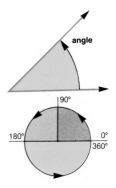

angle

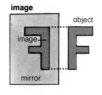

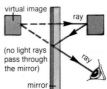

image (*n*) a picture of an object formed by a mirror or a lens (p.58) or formed on the retina (p.126) of the eye, e.g. the image of a person when he looks at himself in a mirror. An image is formed when rays of light, starting from the same point on an object, meet at a point which is the same point on the image as on the object, *see diagram*.

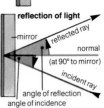

screen (*n*) (1) a flat surface on which a picture can be formed by rays of light, e.g. the screen of a cinema; (2) a thin object, like a wall, which stops light, magnetism (p.69) or any undesirable effect from reaching a place. **screen** (*v*).

real image an image through which rays of light pass, and thus can be put on a screen.

virtual image an image from which rays of light appear to come. As no light rays actually pass through the image, it cannot be put on a screen.

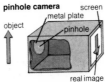

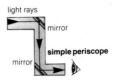

pinhole camera · screen · metal plate · object · pinhole · real image

light rays · mirror · **simple periscope** · mirror

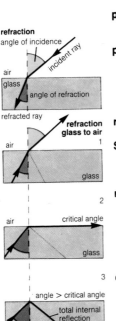

refraction
angle of incidence · incident ray · air · glass · angle of refraction · refracted ray

refraction glass to air
1 · air · glass

2 · air · critical angle · glass

3 · angle > critical angle · total internal reflection · incident ray

pinhole camera a box with a screen at one end and a very small hole in a thin metal plate at the other end. A real image is formed on the screen.

periscope (*n*) an instrument for seeing objects which are above eye-level, e.g. for looking over a wall. A simple periscope uses mirrors; prisms (p.61) can be used instead of mirrors. Lenses can be added so that the periscope also acts as a telescope (p.60).

refraction (*n*) the bending of a ray of light as it passes from one medium to another. **refract** (*v*).

Snell's law for a ray of light that is refracted, the sine of the angle of incidence divided by the sine of the angle of refraction is a constant ratio for any angle of incidence.

refractive index a measure of the ability of a medium (p.52) to refract light. For a particular medium it is equal to the speed of light in a vacuum divided by the speed of light in the medium. It is also measured by: refractive index = (sine angle of incidence) ÷ (sine angle of refraction). In symbols: $n = \sin i / \sin r$.

critical angle when a ray of light passes from one medium to another and the angle of refraction is greater than the angle of incidence (as from glass to air) then the least angle of incidence for which no refraction takes place is the critical angle. At the critical angle, the angle of refraction would be 90°, but the ray is reflected instead by the surface of the medium. For glass to air, the critical angle is 42°.

total internal reflection if the angle of incidence is greater than the critical angle of a medium, total internal reflection takes place.

mirage (*n*) the reflection of light by a layer of very warm air, which has been heated by the Earth. An object and its reflected image are seen; this gives the appearance of a water surface.

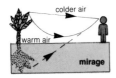

colder air · warm air · **mirage**

lens (*n*) (*lenses*) a piece of glass (or other transparent material) with one or both sides curved, so that it refracts light, and can form an image.

focus (*n*) (1) the point at which rays of light, which have been coming closer together, finally meet; this is a *real focus* as on a screen.
(2) the point from which rays appear to come; this is a *virtual focus*. The *principal focus* of a lens is the point through which parallel rays will pass, or will appear to pass, after being refracted by a lens, *see diagram*.

focal length the distance between the centre of a lens and its principal focus (↑).

converging lens a lens which causes the rays of a parallel beam of light to come closer and pass through a point, i.e. the rays *converge*. It produces a real image at its principal focus. Such a lens is thicker at the centre than at the edge.

diverging lens a lens which causes the rays of a parallel beam of light to spread out and appear to come from a point, i.e. the rays *diverge*. It produces a virtual image at its principal focus. Such a lens is thinner at the centre than at the edge.

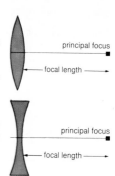

principal focus
focal length

principal focus
focal length

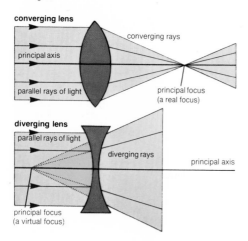

converging lens
converging rays
principal axis
parallel rays of light
principal focus
(a real focus)

diverging lens
parallel rays of light
diverging rays
principal axis
principal focus
(a virtual focus)

convex surface

concave surface

curved mirror

convex (*adj*) describes a surface which curves outwards. A lens with two convex surfaces is a converging lens.

concave (*adj*) describes a surface which curves inwards. A lens with two concave surfaces is a diverging lens.

curved mirror a mirror with either a convex or a concave surface, capable of bringing parallel rays of light to a real image (concave mirror) or a virtual image (convex mirror).

magnify (*v*) to make an image larger than the object. **magnifying** (*adj*).

magnification (*n*) the ratio of the image size to the object size in one direction (one dimension).

principal axis a line passing through the centre of a lens or curved mirror and at a right angle to it.

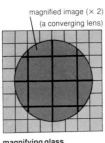

magnified image (× 2)
(a converging lens)

magnifying glass

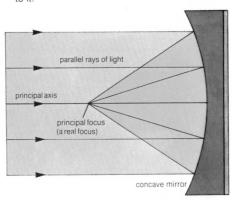

parallel rays of light

principal axis

principal focus
(a real focus)

concave mirror

crystalline lens a transparent structure, a part of the human eye, (p.203) which acts as a converging lens, focusing light onto the retina (p.205) of the eye to form an image. The lens is suspended from a circular muscle which can contract and make the lens more convex; this shortens the focal length so the eye can focus on near objects.

accommodation (*n*) the action of altering the focal length of a lens to see near objects. At rest the eye is focused on distant objects.

defect (*n*) a missing part, property or characteristic, a wrong part or anything that causes a whole to be imperfect, e.g. (a) an eyeball with the wrong shape causes a defect in sight; (b) the loss of elasticity (p.27) in the crystalline lens is a defect of sight causing loss of accommodation (p.59). **defective** (*adj*).

spectacles

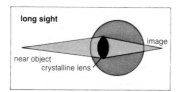

long sight

image
near object
crystalline lens

short sight

parallel rays
distant object
image

long sight a defect in sight caused by the eye focusing near objects behind the retina (p.205). The defect is overcome by using converging lenses in spectacles.

short sight a defect in sight caused by the eye focusing distant objects in front of the retina. The defect is overcome by using diverging lenses in spectacles.

camera (*n*) a device for obtaining photographs. A box has at one end a converging lens which focuses an image on a light-sensitive **film** at the other end. A **shutter**, in front of the lens, prevents light entering the camera until it is opened to take a photograph. The camera can be focused by moving the lens away from the film. A **diaphragm** controls the size of the **aperture** (↓), see diagram.

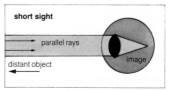

camera
converging lens
diaphragm
shutter
variable distance
light-sensitive film
inside painted black

aperture (*n*) an opening into a space; the size of an opening put in front of a lens.

diaphragm

telescope (*n*) a device for seeing distant objects. Two or more tubes slide into each other; at one end is a converging lens (the objective) and at the other end is another converging lens (the eyepiece). These two lenses form a simple astronomical telescope.

objective (*n*) the lens in a telescope or microscope (↓) nearest to the object.

eyepiece (*n*) the lens in a telescope or microscope (↓) through which a person looks.

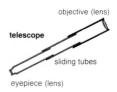

objective (lens)
telescope
sliding tubes
eyepiece (lens)

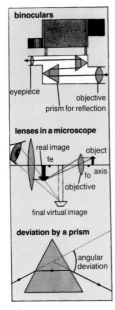

binoculars

eyepiece
objective
prism for reflection

lenses in a microscope

real image
fe
object
fo
axis
objective

final virtual image

deviation by a prism

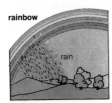

angular
deviation

binoculars (*n*) a device (p.10) which uses two telescopes, one for each eye, for seeing objects at a distance. Prisms (↓), using internal reflection, are used to increase the distance between objective (↑) and eyepiece (↑).

binocular (*adj*) using two eyes for looking at the same object.

microscope (*n*) a device for magnifying (p.59) very small objects. Two converging lenses are used as shown in the diagram. **microscopic** (*adj*).

prism (*n*) a triangular prism is a piece of glass (or other transparent material) with two equal triangular faces joined by three rectangular, flat faces; it is used for refracting or internally reflecting light. **prismatic** (*adj*).

deviation (*n*) a change in direction of the path of an object or a wave motion (p.65), e.g. a deviation of a light ray by a prism. **deviate** (*v*).

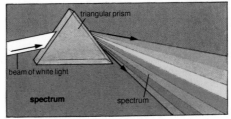

triangular prism

beam of white light

spectrum   spectrum

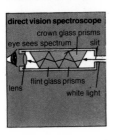

rainbow

rain

direct vision spectroscope

crown glass prisms
eye sees spectrum   slit

flint glass prisms
lens   white light

spectrum (*n*) the result obtained when light is passed through a prism and is split into the following colours: red, orange, yellow, green, blue, indigo, violet. Light from the sun is white light, and it is made up of the colours of the spectrum. **spectral** (*adj*).

rainbow (*n*) the colours of the spectrum formed by refraction (p.57) and reflection (p.56) of light from the sun by raindrops. The effect is seen when a person has his back to the sun, and rain in front of him.

spectroscope (*n*) an instrument for looking at the colours of the spectrum. A simple spectroscope has five glass prisms in a tube (a direct vision spectroscope), *see diagram*.

star (*n*) a luminous (p.55) heavenly body which remains in a definite place relative to other stars. Compare planet (p.55), comet and meteor (↓). The sun is a star.

light year the distance covered by light in 1 year, i.e. approx. 10^{13} km; used in measuring distances of stars from the Earth.

magnitude[2] (*n*) a measure of the brightness of a star. The scale is: first magnitude, second magnitude, etc., with first magnitude as the brightest. A star of one magnitude is approximately 2.51 times brighter than a star of the next lower magnitude.

constellation (*n*) a small group of stars fixed relative to each other; the group is given a name, e.g. the Plough.

nebula (*n*) (*nebulae*) a milky, luminous (p.55) spot in the sky; a great cloud of very hot gas round some stars. New star systems may be formed from nebulae. **nebular** (*adj*).

solar (*adj*) of the sun, e.g. a solar day, the time for the Earth to make one complete turn relative to the sun.

sidereal (*adj*) measured relative to the stars, e.g. in a sidereal day, the Earth makes one complete turn relative to the stars. A sidereal day is about 4 minutes shorter than a solar day.

stellar (*adj*) of a star, e.g. a constellation is a stellar system.

lunar (*adj*) of the moon, e.g. a lunar month is the time from one new moon to the next new moon.

system[1] (*n*) a group of objects or materials which depend on each other and act in agreement with scientific laws to form a whole, e.g. the solar system is a group of planets together with the sun, all obeying the law of gravity and acting on each other, thus forming a whole. **systematic** (*adj*).

galaxy (*n*) a very large group of stars, containing thousands of millions of stars, forming a stellar (↑) system. The solar system is part of a galaxy called the Milky Way. Other galaxies are placed irregularly in space. **galactic** (*adj*).

universe (*n*) the system of all the galaxies. **universal** (*adj*).

constellation

the Plough

ursa major

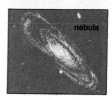

nebula

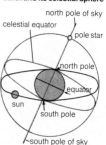

Earth and its celestial sphere

north pole of sky

celestial equator

pole star

north pole

equator

sun

south pole

south pole of sky

comet

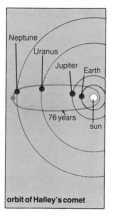

orbit of Halley's comet

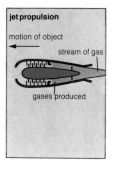

jet propulsion

motion of object

stream of gas

gases produced

comet (*n*) a luminous (p.55) heavenly body with a bright tail, which moves in an orbit (p.54) round the sun. A comet has small mass, but often has great size. The tail points away from the sun. Some comets, e.g. Halley's comet, are regularly seen from Earth, while other comets disappear from the solar system.

meteor (*n*) a solid body (either iron or stone) which enters the Earth's atmosphere and shines brightly because of the heat produced by friction (p.22) with the air; small meteors are completely burned to gases. **meteoric** (*adj*).

meteorite (*n*) a large meteor which is not completely burned to gas, but falls to the Earth.

meteor shower a large number of meteors entering the Earth's atmosphere, mainly seen when the Earth crosses the orbit of a comet.

satellite (*n*) (1) a smaller heavenly body held in orbit (p.54) round a bigger body by gravitational (p.17) attraction, e.g. the moon is a satellite of the Earth. (2) an object, made by man, and put into an orbit by a rocket (↓) is called an artificial satellite.

jet propulsion the pushing of an object in one direction by a stream of gas sent out in the opposite direction. The gas is usually produced by combustion (p.112). The lower the density of the atmosphere, the more efficient is this kind of propulsion.

reaction propulsion a name for jet propulsion.

projectile (*n*) an object pushed or thrown forward in the air by a sudden, great force, so that the object continues in motion when the force no longer acts, e.g. a bullet, fired from a gun, is a projectile. **project** (*v*), **projectile** (*adj*).

rocket (*n*) a projectile driven by jet propulsion (↑). The rocket contains its own propellants (↓) and does not need air for combustion, so it does not depend on the Earth's atmosphere, and can travel in space.

propellant (*n*) a slow explosive, e.g. gunpowder, used to apply force to a projectile. It is a substance which either burns explosively or supplies oxygen for the explosion. Rocket propellants are either solid or liquid substances.

vibrate (*v*), of an elastic (p.27) material, to move regularly, backwards and forwards. An elastic rod, held at one end, vibrates with the free end moving backwards and forwards. When a fluid vibrates, each molecule moves backwards and forwards about a fixed point, because of the fluid's elasticity. **vibration** (*n*), **vibrator** (*n*), **vibratory** (*adj*).

frequency (*n*) the number of times an event is regularly repeated in unit time, e.g. the number of vibrations in 1 second. The symbol for frequency is *f*. **frequent** (*adj*).

period (*n*) the time taken for one complete event in a number of events, when each event takes the same time for completion, e.g. the period of revolution of the Earth round the sun is 1 year.

periodic (*adj*) describes an event which occurs at regular periods. **periodicity** (*n*).

hertz (*n*) the S.I. unit of frequency. A periodic event has a frequency of 1 hertz if its period is 1 second. (*period*) = 1/(*frequency*) when the period is measured in seconds, and the frequency in hertz. The symbol is: Hz.

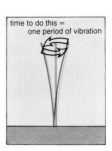

time to do this =
one period of vibration

vibrating rod

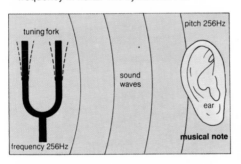

tuning fork

pitch 256Hz

sound waves

ear

frequency 256Hz

musical note

musical note a sound produced by vibrations of regular frequency of the air, e.g. the sound made by a bell and propagated (p.52) by the air.

amplitude (*n*) (1) the distance between the middle and the outer position of a vibrating body. (2) the distance between the middle and the top (or bottom) of a wave (↓).

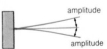

amplitude

amplitude

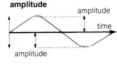

amplitude

amplitude

time

amplitude

pitch (*n*) the place of a note in a musical scale; it is a measure of the frequency of vibration of the source of the note, e.g. a high frequency vibration produces a note of high pitch.

quality (*n*) a property of a musical note which makes it different from another note of the same pitch and loudness. It depends on the overtones (p.67) produced by a musical instrument and enables different instruments to be recognized.

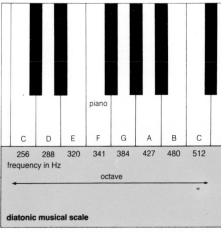

diatonic musical scale

C	D	E	F	G	A	B	C
256	288	320	341	384	427	480	512

frequency in Hz

octave

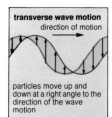

transverse wave motion
direction of motion

particles move up and down at a right angle to the direction of the wave motion

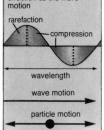

longitudinal wave motion

particles move backwards and forwards in the same direction as the wave motion

rarefaction
compression
wavelength
wave motion
particle motion

musical scale musical notes of increasing pitch at regular intervals (p.68) form a musical scale.

wave motion the sending of energy by a periodic (↑) movement in a medium (p.52) in the form of a wave, *see diagram*. In a material medium, the particles move about a central position and produce the wave motion.

transverse wave a wave in which particles of a material medium move up and down, about a central position, at a right angle to the direction of the wave motion.

longitudinal wave a wave in which particles of a material medium move backwards and forwards, about a central position, in the same direction as the wave motion. Sound waves are longitudinal waves.

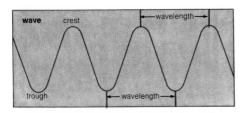

wavelength (*n*) the distance between a point in a wave and the next point at the corresponding place moving in the same direction, i.e. the distance between one crest and the next crest, *see diagram*, or between two troughs. The symbol for wavelength is λ. The wavelength is equal to the speed of the wave motion (*v*) divided by its frequency (*f*). The frequency of a wave motion is the number of crests (or troughs) that pass a point in 1 second. In symbols: $v = f\lambda$.

sound wave a longitudinal wave produced by a vibrating object. As the object vibrates, it sends out: (1) a wave of high pressure, known as a **compression**; (2) a wave of low pressure, a **rarefaction**. In between, in air, the pressure returns to atmospheric pressure. Sound needs a material medium for its propagation (p.52). A sound wave has a speed of about 330 m/s in air. The wavelength of a sound wave is the distance between one compression and the next.

natural frequency the frequency of vibration of a solid object, or a column of fluid, when free to vibrate and not acted upon by an outside force.

resonance (*n*) the cause, in an object, of very large amplitudes (p.64) of vibration which are produced when a periodic (p.64) force, with the same frequency as the natural frequency of the object, is applied to the object. The object is in resonance with the force, e.g. a loud note of the same frequency as that of a wine glass can break the glass by resonance. **resonant** (*adj*).

echo (*n*) the effect produced by the reflection of sound from a surface, e.g. an echo from the sea bed, used in an echo sounder, *see diagram*.

echo sounding

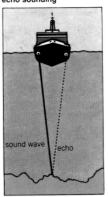

stationary wave the kind of wave formed when a stretched string vibrates, *see diagram*. The waves do not move along the string; they are caused by two waves, of the same frequency, travelling in opposite directions and opposing each other's motion. The wavelength (↑) of a stationary wave is twice the distance between two nodes (↓), or between two antinodes (↓). Stationary waves are also produced in vibrating columns of air.

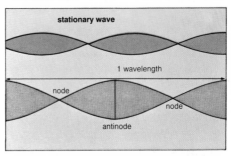

stationary wave

1 wavelength

node

node

antinode

node[1] (*n*) a point in a stationary wave where there is no amplitude (p.64) of vibration. **nodal** (*adj*).

antinode (*n*) a point in a stationary wave where there is the biggest amplitude of vibration. An antinode is midway between two nodes.

fundamental frequency the lowest frequency of vibration of a stationary wave. For a stretched string the fundamental frequency of vibration is produced when the wavelength is twice the length of the string. For a wind instrument, the fundamental frequency is produced when the wavelength is four times the length of the vibrating column of air. When vibrating with its fundamental frequency alone, a musical instrument gives out a **pure note**.

overtone (*n*) a note of higher frequency than the pure note (↑) given out by a musical instrument. A stretched string can produce overtones as shown on page 68. Most musical instruments produce overtones, and the addition of these overtones forms the quality (p.65) of the note.

harmonic (*n*) an overtone (p.67) with a frequency
that is equal to the fundamental frequency
multiplied by a whole number, e.g. the second
harmonic has a frequency twice that of the
fundamental frequency, the third harmonic,
three times that of the fundamental frequency.
Harmonics can be added to a wave which has a
single frequency to produce a **complex** wave.

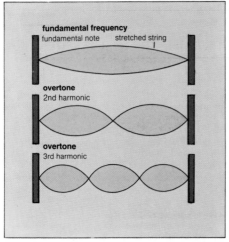

interval (*n*) (1) a space in distance between two
objects, e.g. an interval of 3 metres between
posts in a line, or a space in time between two
events, e.g. the interval between meals. (2)
the difference between two points on a scale.
The interval between the pitch of two musical
notes is measured by the ratio of their
frequencies, e.g. two notes with pitches of
288 Hz and 256 Hz have a frequency interval
of 9:8.

octave (*n*) (1) the interval between any two
frequencies having a ratio of 2:1, e.g. notes with
pitches of 512 Hz and 256 Hz are an octave
apart *see diagram* (p.65). (2) in any wave
motion, the band of frequencies between any
two frequencies which are in the ratio of 2:1.

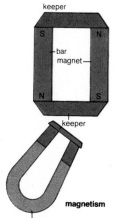

keeper

bar magnet

keeper

magnetism

horseshoe magnet

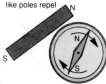

like poles repel

unlike poles attract

magnet (*n*) a solid object that attracts iron and attracts or repels (↓) other magnets. When free to turn, it points in a north-south direction. It possesses the property of **magnetism** (*n*). **magnetic** (*adj*), **magnetize** (*v*).

permanent magnet a magnet that does not lose its magnetism; usually made of steel, or an alloy (p.103) of steel.

temporary magnet a piece of iron, or other magnetic material, that is a magnet only as long as it is influenced by magnetizing force; soft iron is often used for temporary magnets.

pole (*n*) points near each end of a magnet at which the magnetism appears to be strongest. One end is called a north pole because it points north when the magnet is free to turn; the other end is a south pole. **polar** (*adj*).

keeper (*n*) a piece of soft iron put against the north pole of one magnet and the south pole of the same magnet or another magnet; it prevents any tendency (p.15) of a permanent magnet to lose its magnetism.

like (*adj*) (1) two north poles or two south poles are called like poles. (2) two positive charges (p.71) or two negative charges are called like charges. (3) same in nature, or character.

unlike (*adj*) (1) a north pole and a south pole are unlike poles. (2) a positive charge and a negative charge are unlike charges.

repel (*v*) to make an object, or anything else, go away, e.g. like magnetic poles repel each other and unlike magnetic poles attract each other.

magnetic field the space round a magnet or an electric current in which a magnetic material experiences a magnetic force of attraction, or a magnet sets in the direction of the magnetic force from the magnet.

magnetic field

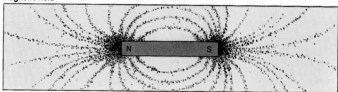

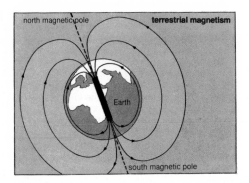

terrestrial magnetism the magnetism of the
Earth; it exerts (p.23) a magnetic field covering
the Earth, similar to the field that a powerful
magnet, at the centre of the Earth, would
produce.

compass (*n*) an instrument with a magnet, free
to turn on a pivot, in a case, with the directions
north, south, east and west marked on a scale.
In a **mariner's compass**, two or more magnetic
needles are fixed to a card which floats in
alcohol and water. The compass is used to
find directions on the Earth.

dip circle an instrument with a magnetic needle,
free to turn in a vertical plane, which is pulled
down (i.e. dips) at one end by the Earth's
magnetic field.

magnetic variation the angle between magnetic
north and geographical north at any point on the
Earth. The north pole of the Earth's magnetic
field is not in the same place as the North
Pole of the Earth. Magnetic variation lies
between 30°W to 30°E, depending on the
position of the compass on the Earth.

magnetic declination another name for magnetic
variation.

magnetic dip the angle between the horizontal
and the direction of the Earth's magnetic field.
It varies between 0° about the equator to 90° at
the Earth's magnetic poles.

magnetic inclination a name for magnetic dip.

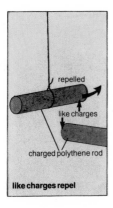

like charges

repelled

like charges

charged polythene rod

like charges repel

discharge of a cloud

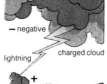

— negative

lightning

charged cloud

+ positive

Earth

charge (*n*) (1) the correct quantity of a material that is put in, or is needed, by a device, e.g. a charge of explosives used in a mine. (2) a charge of electricity cannot be described; its effects are: (a) an attraction for an unlike (p.69) charge or repulsion for a like charge; (b) the production of a spark when the charge escapes to earth, i.e. discharged; (c) the forming of an electric field. Symbol for electric charge: *Q*. **charged** (*adj*), **charge** (*v*), **discharge** (*v*), **discharge** (*n*).

electrostatic (*adj*) describes effects caused by electric charges at rest, such as an electric charge on an object.

spark (*n*) an instantaneous (p.21) appearance of light and sound in an electrostatic discharge of short duration.

lightning (*n*) a very large spark caused by the discharge of a cloud carrying an electrostatic charge.

thunder (*n*) the sound produced by the discharge of a charged cloud.

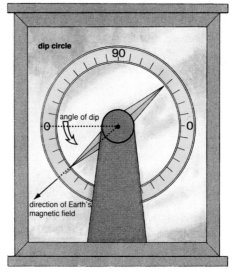

dip circle

90

angle of dip

0

0

direction of Earth's magnetic field

primary cell a device which produces a flow of electric charge, i.e. an electric current, by means of a chemical reaction. The simplest cell has two pieces of different metals, in acid, or alkali, contained in a vessel.

local action an effect caused by impurities (p.99) in the zinc plate of a primary cell. Bubbles of hydrogen are formed on the plate and the efficiency of the cell is made less. Rubbing the zinc with mercury prevents local action.

polarization (*n*) an effect produced when a simple primary cell produces electric current. Bubbles of hydrogen form on the copper or positive plate and the electric current quickly falls to a small value. All primary cells suffer from polarization. **polarize** (*v*).

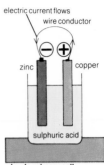

simple primary cell

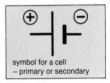

symbol for a cell
– primary or secondary

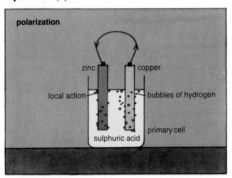

depolarization (*n*) the action to prevent polarization. A substance is added which chemically combines (p.94) with the hydrogen formed; such a substance is called a **depolarizer** (*n*). **depolarize** (*v*).

Leclanché cell a primary cell using a zinc rod and a carbon rod. The carbon rod has manganese (IV) oxide (manganese dioxide) powder around it, and is contained in a porous pot. The zinc rod and the porous pot stand in a solution (p.89) of ammonium chloride. The zinc rod is negative and the carbon rod is positive. The manganese (IV) oxide is a depolarizer (↑).

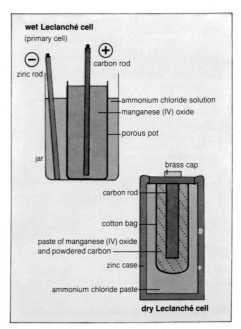

wet Leclanché cell
(primary cell)

⊖ zinc rod

⊕ carbon rod

ammonium chloride solution

manganese (IV) oxide

porous pot

jar

brass cap

carbon rod

cotton bag

paste of manganese (IV) oxide and powdered carbon

zinc case

ammonium chloride paste

dry Leclanché cell

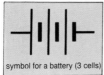

symbol for a battery (3 cells)

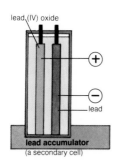

lead (IV) oxide

lead

lead accumulator
(a secondary cell)

battery (*n*) a number of primary or secondary (↓) cells working together.

dry battery one or more dry cells working together. A dry cell is a Leclanché cell using ammonium chloride in the form of a paste and not a solution.

secondary cell an electric cell which produces an electric current by a chemical reaction. After being discharged (p.71) the cell can be recharged by passing an electric current through it in the opposite direction. A primary cell cannot be used in this way.

storage cell another name for secondary cell.

accumulator (*n*) a storage cell, or a battery of storage cells. The commonest kind is the lead-acid accumulator, with a negative plate of lead and a positive plate of lead (IV) oxide (lead dioxide) in sulphuric acid.

electrical (*adj*) concerning electricity, but not always carrying an electric current, e.g. electrical engineer.

electric (*adj*) describes any device which uses or produces a charge of electricity or any effect of such charge, e.g. an electric cell, an electric bell, an electric field.

current (*n*) (1) a fluid moving in a particular direction, or the motion of the fluid. (2) the flow of electric charge through a solid or a liquid. An electric current is measured in amperes. The symbol is: *I*.

conductor (*n*) a material through which an electric current or heat can flow. All metals are good conductors of both electric charge and heat. **conduct** (*v*).

insulator (*n*) a material or an object which prevents the flow of an electric current or heat. Most non-metals and all gases are insulators, e.g. sulphur and air are insulators.

insulate (*v*) to prevent electric current or heat escaping from where they are used and to prevent them from entering where they are not wanted. **insulation** (*n*) (1) the action of insulating. (2) material used for insulating.

electromotive force the force which drives an electric current round an electrical circuit (p.76). An electric cell (↑) or a generator (p.80) produce an electromotive force; for any one kind of cell, its electromotive force is constant if there is no polarization (p.72). Abbreviation: e.m.f.; symbol: E. An e.m.f. is measured in volts (↓).

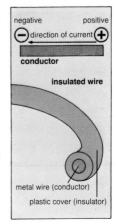

negative   positive

direction of current

conductor

insulated wire

metal wire (conductor)

plastic cover (insulator)

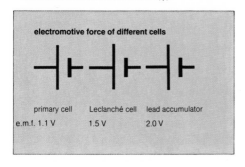

electromotive force of different cells

primary cell	Leclanché cell	lead accumulator
e.m.f. 1.1 V	1.5 V	2.0 V

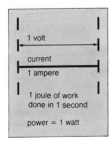

**relation between volt,
ampere, joule and watt**

potential difference the difference in electric
force between any two points in an electrical
circuit (p.76). Abbreviation: p.d.; symbol: *V*.
An electric current always flows from a higher to
a lower p.d.; it flows from a positive to a
negative p.d. (This is a convention).

resistance[2] (*n*) the force which opposes the flow
of an electric current through a conductor (↑).
The resistance of an object, e.g. a wire, is
measured in ohms (↓). Symbol: *R*. (See resistor
p.76).

ampere (*n*) the S.I. unit of electric current. If
a current of 1 ampere flows through each of
two parallel conductors (↑), put 1 metre apart in
a vacuum, then there will be a force of 2×10^{-7}
newtons per metre of length of the conductors.
Symbol: A.

coulomb (*n*) the S.I. unit of electric charge; the
quantity of charge transferred by 1 ampere
in 1 second between any two points of a
circuit. Symbol: C. A current of 6A flowing for
3 seconds transfers 18C.

volt (*n*) the S.I. unit of electric potential; 1 volt is
the difference of potential between two points
if 1 joule of work is done, transferring 1 coulomb
of charge between the points. **voltage** (*n*) a
potential difference measured in volts.

ohm (*n*) the S.I. unit of resistance; a resistance
of 1 ohm exists between two points in a circuit
(p.76) when a potential difference of 1 volt
between the points produces a current of
1 ampere. Symbol: Ω.

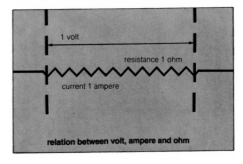

relation between volt, ampere and ohm

circuit (*n*) a continuous path of conductors and other electrical devices along which an electric current can flow; this is a *closed circuit*. If the circuit is broken at a point, so the current cannot flow, it is an *open circuit*.

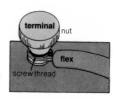

terminal (*n*) a metal nut on a screw thread, to which a wire can be connected by tightening the nut. It joins wires to resistors (↓), cells, and other electric devices.

flex (*n*) a wire that bends to any shape to make electrical connections, i.e. to join parts of a circuit. It is always covered with insulating material.

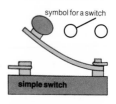

switch (*n*) a device which is used to join (switch on) parts of a circuit and to break (switch off) parts of a circuit. The most common switch is for connecting and disconnecting an electric light. **switch on** (*v*), **switch off** (*v*).

open circuit see circuit (↑).

resistor (*n*) a device which offers resistance to an electric current. If the value of the resistor is known in ohms (p.75) and is constant, it is a *fixed resistor*.

variable resistor a resistor whose resistance can be changed. A contact, *see diagram*, slides round a length of wire, and thus varies the resistance.

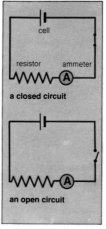

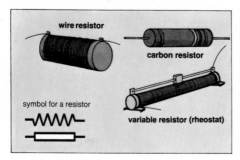

rheostat (*n*) another name for a variable resistor.

heating element a part of a heating appliance; it is a length of wire, of low resistance, that is heated when an electric current flows through it. It is used in electric fires, kettles, stoves, irons.

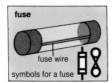

fuse

fuse wire

symbols for a fuse

filament[1] (*n*) a piece of very thin wire of high resistance which, when an electric current passes through it, becomes very hot and gives out light; used in an electric light bulb. A filament is also used in a thermionic valve (p.85) to give out electrons (p.106).

fuse (*n*) a device containing a piece of wire which melts if too great an electric current is passed through it. This breaks the circuit in which the fuse is placed; the fuse acts as a safety device.

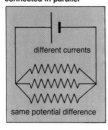

connected in parallel

different currents

same potential difference

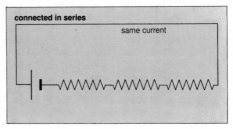

connected in series

same current

series circuit a circuit in which all the parts are connected one after the other so that the same electric current flows through each part.

parallel circuit a circuit in which all the parts are connected to the same electromotive force (p.74) so that the same potential difference (p.75) is applied to each part.

Ohm's law the electric current flowing through a metallic conductor is proportional to the potential difference between the ends of the conductor, if the temperature is kept constant. The ratio of potential difference divided by current is the resistance of the conductor. In symbols: $I \propto V$; $V = IR$, where V is in volts, I in amperes, and R in ohms.

galvanometer (*n*) an instrument for measuring electric currents, particularly small currents. It has a scale so that currents can be compared, but it does not measure current in amperes.

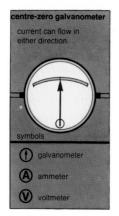

centre-zero galvanometer

current can flow in either direction

symbols

galvanometer

ammeter

voltmeter

ammeter (*n*) an instrument for measuring electric current in amperes.

voltmeter (*n*) an instrument for measuring electromotive force or potential difference (p.75) in volts.

Seebeck effect if two wires of different metals are joined at their ends to form a closed circuit (p.76) with two **junctions**, and if the two junctions are kept at different temperatures, then an electric current flows in the circuit.

thermoelectric (*adj*) describes the production of electric current directly from heat, e.g. *thermoelectric couple*: two wires of different metals used in the Seebeck effect (↑); *thermoelectric junction*: the join between two wires in a thermoelectric couple.

thermocouple (*n*) another name for thermo-electric couple (↑).

thermopile (*n*) an instrument with thermocouples (↑) connected in series behind a cone. It detects heat radiation and produces a thermoelectric current, measured by a galvanometer.

Peltier effect if two wires of different metals are joined and an electric current is passed through the junction, the junction is either warmed or cooled depending on the direction of the flow of current.

electromagnetism (*n*) (1) the study of magnetic effects produced by electric currents and of electric currents produced by magnetic fields. (2) magnetism produced by an electric current. **electromagnetic** (*adj*).

coil (*n*) a wire wound round a solid object (a former) in rings. The length of a coil is usually much shorter than its diameter. Coils are used in circuits for their electromagnetic effects.

winding (*n*) wire wound round part of an electrical device so that it acts as a coil, e.g. the winding on an armature (p.80).

turns (*n.pl.*) the rings of wire in a coil, e.g. a coil of 20 turns has 20 complete circles of wire wound round the former, *see diagram*.

core (*n*) (1) the middle part of a solid object. (2) the middle part of a coil or solenoid (↓), e.g. an iron rod in the middle of a solenoid; if a coil has no metal core it has an air core.

solenoid (*n*) a kind of coil in which the length is very much greater than the diameter. When an electric current passes through it, the solenoid acts as a magnet.

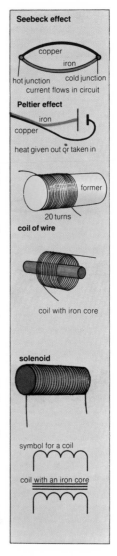

Seebeck effect

copper

iron

hot junction   cold junction

current flows in circuit

Peltier effect

iron

copper

heat given out or taken in

former

20 turns

coil of wire

coil with iron core

solenoid

symbol for a coil

coil with an iron core

electromagnet

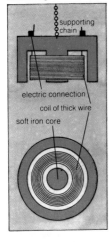

supporting chain

electric connection

coil of thick wire

soft iron core

electromagnet (*n*) a coil with a soft iron core; when an electric current flows through the coil it acts as a magnet; when the current stops flowing, it loses its magnetism.

microphone (*n*) a device for transforming sound waves into an electric current. A diaphragm, *see diagram*, vibrates to sound waves, and pushes against carbon granules; the resistance of the granules varies with pressure, and causes a variation in current passing from the diaphragm to the iron plate. This varying current is passed to a receiver.

receiver (*n*) a device for transforming electric current into sound waves. A diaphragm is held above a permanent magnet; coils of wire are wound round the magnet, *see diagram*. Electric current (passed from the microphone) flows in the coils and the varying current causes the diaphragm to vibrate and give out sound waves.

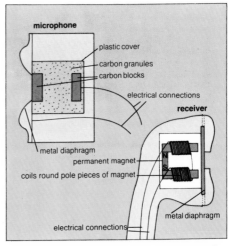

microphone

plastic cover

carbon granules

carbon blocks

electrical connections

receiver

metal diaphragm

permanent magnet

coils round pole pieces of magnet

metal diaphragm

electrical connections

simple relay

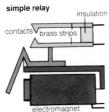

insulation

contacts

brass strips

electromagnet

relay (*n*) an electrical device which uses a small electric current to control a greater current in another circuit by switching (p.76) it on or off. Electrical relays use electromagnets to work a switch, *see diagram*.

telephone (*n*) a circuit connecting a microphone, a receiver and a source (p.52) of electric energy, by wires. **telephonic** (*adj*), **telephony** (*n*).

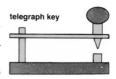

telegraph key

telegraph (*n*) a device for sending messages over a distance by an intermittent (p.81) current conducted by wires. A key, *see diagram*, makes the intermittent current which is heard in a receiver. A code represents letters and numbers; it is called the Morse Code.

moving-coil (*adj*) describes an instrument which uses the motion of a coil in a magnetic field. When a current is passed through the coil, it acts as a magnet and turns to set in the direction of the magnetic field.

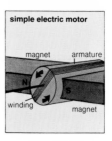

simple electric motor

magnet   armature

winding   magnet

motor[1] (*n*) a device, other than an engine, which produces motion. An electric motor consists of a coil of wire wound round an armature (↓) which turns between the poles of a magnet.

armature (*n*) a piece of soft iron, turning on an axle, with a coil wound round it. When an electric current passes through the coil, the armature turns, producing mechanical motion. See generator (↓).

generator (*n*) a machine for transforming mechanical energy into electrical energy. The simplest generator has an armature, with a coil wound on it, turning inside the poles of a permanent magnet. When the armature is turned, a current flows in the coil. Larger generators use electromagnets with stators and rotors (↓); they provide electricity for houses.

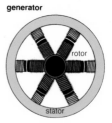

generator

rotor

stator

dynamo (*n*) another name for generator, especially a small generator supplying direct current (↓).

rotor (*n*) the part of a generator, or turbine (p.44), that turns.

stator (*n*) the part of a generator, or turbine, that remains stationary.

direct current an electric current that flows in one direction only. Abbreviation: d.c.

alternating current an electric current which increases to a definite value, then decreases, finally changing direction and reaching the same value in the opposite direction, then increases again and repeats the changes. The frequency

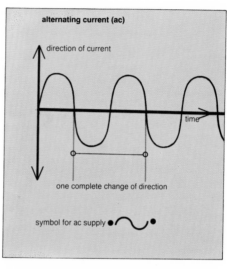

alternating current (ac)

direction of current

time

one complete change of direction

symbol for ac supply

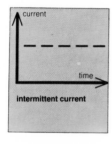

intermittent current

of the current is the number of complete
changes made in one second. Abbreviation: a.c.
Alternating current is produced by a generator.

rectifier (*n*) an electrical device which converts
alternating current to direct current. Thermionic
valves (p.85) are often used for this purpose.
rectified (*adj*), **rectify** (*v*).

intermittent (*adj*) describes an action or an effect
that starts and stops time after time, e.g. the
ringing sound of a telephone is intermittent.

fluctuating (*adj*) describes the values of a physical
quantity (p.13) which vary above and below
an expected value, e.g. the sound of a radio can
fluctuate above and below the expected
loudness.

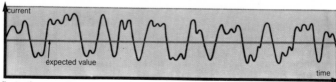

fluctuating current

induced current if a conductor, e.g. a coil, is
moved in a magnetic field, or there is a change
in a magnetic field round a coil, then an
induced current flows in the coil.

Faraday's law whenever there is a change in the
magnetic field passing through a conductor an
electromotive force is induced; the strength of
the e.m.f. is proportional to the rate of change
of the strength of the magnetic field.

Lenz's law the direction of the induced current
is always such that it opposes the change
producing it, e.g. if the north pole of a magnet
enters a coil, the induced current in the coil
makes the face of the coil a north pole, i.e.
opposing entry of the magnet.

transformer (*n*) an apparatus which changes the
voltage of alternating current (p.80). It consists of
two coils of wire wound round the same iron
core, *see diagram*. One coil contains a few turns
of thick wire and the other coil many turns of
thin wire. An alternating current is passed into
one coil (the primary) and an induced alternating
current, of the same frequency, is obtained from
the other coil (the secondary).

turns ratio the ratio of the number of turns in
the secondary winding to the number of turns in
the primary winding of a transformer, e.g. 5000
secondary turns, 100 primary turns; turns ratio =
5000/100 = 50.

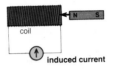
induced current

simple transformer

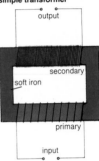

transformer

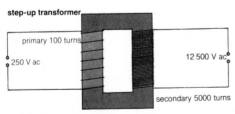
step-up transformer

step-up (*adj*) giving increased voltage, e.g. a
step-up transformer has a turns ratio (↑) greater
than 1.

step-down (*adj*) giving decreased voltage, e.g. a
step-down transformer has a turns ratio (↑) less
than 1.

secondary coil with
primary coil underneath
soft iron

symbol for a transformer

discharge tube

cathode rays
(stream of electrons)

cathode

very high
voltage

anode

glass tube

very low pressure

discharge tube a glass vessel with electric current passing from anode to cathode in a high vacuum.

radiation[2] (*n*) the spreading of energy by electromagnetic waves, i.e. light, radiant heat, X-rays (p.84), radio and gamma rays (p.108). **radiate** (*v*).

infra-red rays electromagnetic waves with wavelengths longer than those of red light in the visible spectrum. These are the rays of radiant heat; they pass through mist and some solid substances which are opaque (p.53) to light rays.

ultra-violet rays electromagnetic waves with wavelengths shorter than those of violet light in the spectrum. These rays cannot be seen, but they act on photographic film, and cause burning of the skin when in sunshine. Abbreviation: u.v. rays; u.v. light.

cathode rays a stream of electrons (p.106) from the cathode (p.100) of a cathode-ray tube (p.86).

bombard (*v*) to send a stream of particles, either charged or not, with force against a surface, e.g. a stream of electrons bombards the screen of a television set (p.86). **bombardment** (*n*).

target (*n*) any surface at which a stream of particles is aimed or directed.

electromagnetic waves

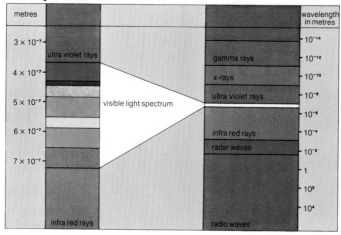

metres				wavelength in metres
3×10^{-7}	ultra violet rays			10^{-14}
			gamma rays	10^{-12}
4×10^{-7}			x-rays	10^{-10}
			ultra violet rays	10^{-8}
5×10^{-7}		visible light spectrum		10^{-6}
6×10^{-7}			infra red rays	10^{-4}
			radar waves	10^{-2}
7×10^{-7}				1
				10^{2}
				10^{4}
	infra red rays		radio waves	

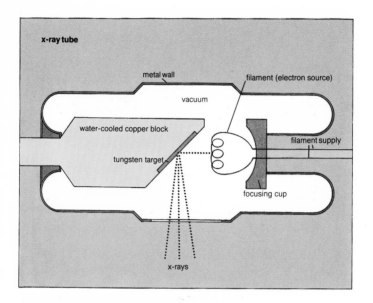

X-ray tube an apparatus for producing X-rays,
see diagram. A heated filament gives out a
stream of electrons; these are focused, by a cup,
onto a target. The filament is the cathode
(negative electrode) and the target is in a
copper block which is the anode (positive
electrode). The electrons have a negative
charge and are accelerated by repulsion from
the cathode. When the electrons hit the target,
X-rays are produced, together with a large
quantity of heat.

X-rays electromagnetic waves with a very short
wavelength, between approximately 10^{-9} and
10^{-11} of a metre. The greater the potential
difference between anode and cathode of an
X-ray tube, the shorter the wavelength of the
X-rays. X-rays pass through many materials
opaque to light, e.g. flesh and bones, with bones
absorbing more than flesh. The shorter the
wavelength of an X-ray, the more easily it passes
through a material.

diode valve

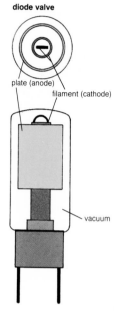

plate (anode)

filament (cathode)

vacuum

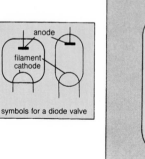

anode

filament
cathode

symbols for a diode valve

valve² (*n*) an electrical device which allows electric current to pass through in one direction only. Valves are called diode, triode, pentode, etc. according to the number of electrodes (↓) they have.

thermionic valve another name for valve² (↑).

electron tube another name for valve² (↑).

tube (*n*) another name for valve², used in America.

electrode¹ (*n*) a wire, rod, or plate conducting electric current into, or out of, any device.

diode (*n*) a valve² (↑) with two electrodes. The cathode (negative electrode) is a heated filament, which gives off electrons (p.106). The anode (positive electrode) is a plate. The anode attracts electrons, and an electron current flows from cathode to anode. This is a flow of negative charge, and is opposite to the usual way of saying current flows from positive to negative. If an alternating voltage is applied to the anode and cathode, then current flows in one direction only and the alternating current is rectified (p.81). This is the main use of diodes.

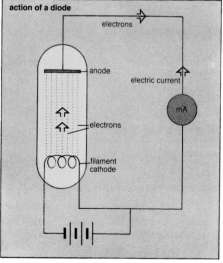

action of a diode

electrons

anode

electric current

electrons

mA

filament
cathode

triode (*n*) a valve[2] (p.85) with three electrodes; a grid (a wire net) is put between the anode and cathode. The grid is given a negative voltage, and a small change in this voltage has a big effect on the electron current, and gives a big voltage at the anode. The triode thus steps-up (p.82) voltage.

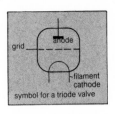

grid — anode

filament
cathode

symbol for a triode valve

cathode-ray tube (*see diagram*). A heated filament gives out electrons which are focused into a narrow stream, accelerated by the anode, and then pass between two sets of parallel plates. The electrons produce light when they hit the screen. The X and Y parallel plates can make the electrons hit any part of the screen because they are given an electric charge which repels or attracts the electrons.

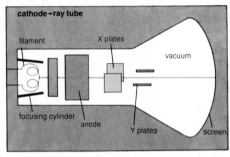

cathode–ray tube

filament   X plates

vacuum

focusing cylinder
anode   Y plates   screen

oscilloscope (*n*) an instrument using a cathode-ray tube which produces an image on the screen of varying electrical voltages. It is also used in radar (↓) to show the position of objects.

radio (*n*) the use of electromagnetic waves to send messages from a microphone in a radio station to a receiver in a radio set.

television (*n*) the use of radio waves to send pictures to a television set. The set has a cathode-ray tube which builds the picture from 625 lines (or 405 lines) each of 400 small spots of light.

radar (*n*) an abbreviation of Radio Detection And Ranging. Electromagnetic waves, of wavelengths of a few centimetres, are reflected from distant objects and the reflections are recorded on the screen of a cathode-ray tube.

Chemistry

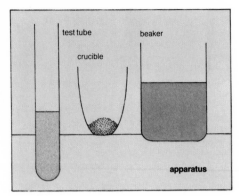

apparatus

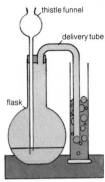

apparatus

apparatus (*n*) any instrument, or collection of instruments, or objects such as test-tubes, beakers, flasks, pipettes, put together for work in science.

test-tube (*n*) a tube, closed at one end used to make chemical tests.

beaker (*n*) a vessel used to hold liquids or solutions, and to heat them.

crucible (*n*) a small cup used for heating solids strongly.

flask (*n*) a vessel used for boiling liquids, preparing gases.

Woulfe-bottle a vessel used for preparing gases.

funnel (*n*) a device used for filtering solutions.

thistle funnel a funnel used for adding liquids to flasks.

delivery tube a tube used for conducting gases.

gas-jar (*n*) a vessel used for collecting gases.

pneumatic trough a vessel used with a gas-jar for collecting gases.

eudiometer tube a glass tube used for experiments on gases.

Kipp's apparatus an apparatus used for supplying a gas.

burette (*n*) a glass tube used for measuring liquids to 0.1 cm^3.

pipette (*n*) a piece of apparatus used for measuring a fixed volume of liquid, usually 10 cm^3, 25 cm^3 or 50 cm^3.

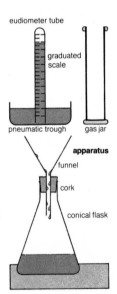

apparatus

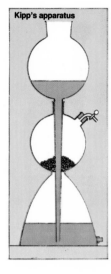

Kipp's apparatus

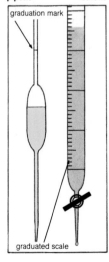

pipette   burette

graduation mark

graduated scale

dissolve (v) to make a solid or a gas disappear, or for it to disappear, into a liquid, e.g. to dissolve sugar in coffee, to dissolve common salt in water. Compare **melt**: when sugar is heated, it melts, when added to water, it dissolves without heating.

solute (n) any solid or gas which, when added to water, or other liquid, will dissolve, e.g. (a) when common salt is added to water, the salt is the solute; (b) when carbon dioxide gas is added to water to make soda water, the gas is the solute.

solvent (n) the liquid in which a solute is dissolved, e.g. (a) water is the solvent and common salt is the solute when salt dissolves in water; (b) paint is the solute and turpentine is the solvent when paint is dissolved in turpentine.

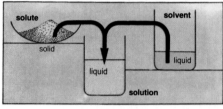

solution (n) the result of dissolving a solute in a solvent, e.g. salt solution (usually called brine) is formed when common salt is dissolved in water. Unless described in some other way a solution is always in water.

saturated (adj) of a solid, cannot absorb (p.163) more liquid; of a solution, cannot dissolve more solute; of a gas, cannot contain more vapour, e.g. (a) when the air is saturated it cannot contain more water vapour, so water cannot evaporate into the air; (b) a saturated solution of common salt cannot dissolve more salt.

unsaturated (adj) of a solution, it can dissolve more solute; of a gas, it can contain more vapour, e.g. (a) unsaturated air can contain more water vapour; (b) an unsaturated solution can dissolve more solute.

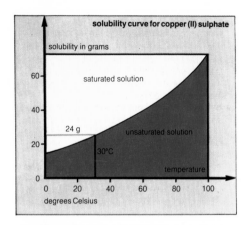

solubility (*n*) the mass of a solute in grammes
that can be dissolved in 100 g of a solvent
to form a saturated solution at a given
temperature. Water is considered the usual
solvent, any other solvent must be named, e.g.
the solubility of copper (II) sulphate is 24 g at
30°C, so 24 g of the solute dissolved in 100 g
of water at 30°C forms a saturated solution.
soluble (*adj*).

insoluble (*adj*) describes a solid or a gas which
does not dissolve in a named solvent, e.g.
hydrogen (a gas) is insoluble in water.

concentration (*n*) (1) a measure of the amount of
solute, dissolved in a particular volume of its
solution. The measurement can be (a) grammes
per dm³; (b) moles (p.105) per dm³; (c) cm³
of gas per cm³ of solution, e.g. a concentration
of 125 g of copper (II) sulphate in 1 dm³ of
solution. (2) the action of increasing the
concentration of a solution by evaporating (p.42)
the solvent with heat; the result is a more
concentrated solution. **concentrate** (*v*),
concentrated (*adj*).

dilute (*v*) to make a solution less concentrated
by adding more solvent. **dilution** (*n*).

dilute (*adj*) describes a solution which has been
diluted, or one which has a low concentration.

immiscible liquids

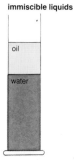

miscible (*adj*) describes a liquid which will mix with another named liquid, e.g. alcohol is miscible in water.

immiscible (*adj*) describes a liquid which will not mix with another named liquid, e.g. olive oil is immiscible in water; it forms two layers of liquid with the oil on top of the water.

separate (*v*) (1) of immiscible liquids, to form two layers. (2) to obtain each substance from a mixture (p.95). **separation** (*n*), **separate** (*adj*), **separable** (*adj*).

precipitate (*n*) a solid thrown out of solution (p.89), when one solution is added to another solution or a gas is passed into a solution, because the substance formed is insoluble, e.g. (a) adding silver nitrate solution to sodium chloride solution forms a precipitate of insoluble silver chloride; (b) passing carbon dioxide gas into lime water forms a precipitate of insoluble calcium carbonate. **precipitate** (*v*).

precipitation[2] (*n*) the forming of a precipitate (↑).

precipitation

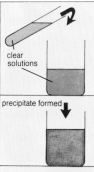

clear solutions

precipitate formed

filter (*v*) to remove solid material from a mixture of liquid and solid, e.g. to filter off a precipitate from a mixture of liquid and precipitate. **filter** (*n*).

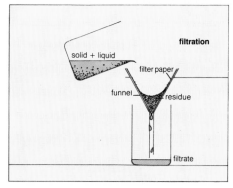

filtration

solid + liquid

filter paper

funnel — residue

filtrate

filtrate (*n*) the liquid part which has passed through a filter.

residue (*n*) the solid material separated when a mixture of liquid and solid is filtered.

evolve (v) (1) to form a gas as a result of a
chemical reaction (p.96) with the gas escaping
into the atmosphere, e.g. when zinc is added to
sulphuric acid, hydrogen is evolved, i.e. formed,
and escapes from the solution into the air. (2)
to set free heat, or sparks. **evolution** (n).

collect (v) (1) to gather, or to come, together
in one place, e.g. to collect different kinds of
rocks. (2) to lead a gas into a vessel, e.g. to
collected hydrogen in a gas-jar. **collection** (n).

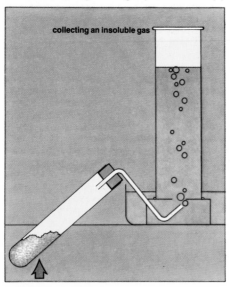

collecting an insoluble gas

bubble (n) a ball of gas either in a liquid, or
with a thin skin of liquid round it, e.g. the
bubbles in soda water.

bubble (v) to pass bubbles of a gas through a
liquid, e.g. to bubble carbon dioxide gas through
lime water.

sublime (v) to change a solid substance to a
vapour by heat, and then change the vapour
back to a solid by cooling it, e.g. ammonium
chloride vaporizes on heating, and sublimes
back to a solid on cooling.

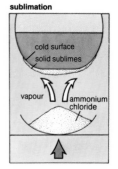

sublimation

cold surface
solid sublimes

vapour ammonium
chloride

experiment (*n*) an exercise, using apparatus (p.88) or instruments (p.10), to observe the behaviour of objects, materials, or organisms (p.147) under controlled conditions, e.g. (a) an experiment to find the refractive index of glass; (b) an experiment to see how chalk behaves when heated. **experiment with** (*v*).

preparation (*n*) the making and collecting of a quantity of a substance using a chemical reaction (p.96), e.g. the preparation of oxygen by heating potassium chlorate, and collecting the gas in a gas-jar.

investigate (*v*) to carry out, for a particular purpose, an experiment (↑) on an object or a material and to record carefully all observations (↓), e.g. to investigate the chemical and physical properties of carbon.

chemical property a property (p.27) which describes the way in which a substance acts when heated, electrolysed (p.100), or added to other substances so that a chemical change (p.94) takes place. All the chemical properties of a substance make up its chemical nature.

test (*n*) a simple exercise carried out: (1) to see if apparatus or instruments are working correctly; (2) to find out whether a particular substance is present or absent. **test** (*v*).

observation (*n*) the intentional use of seeing, hearing, smelling, tasting, or touching, using knowledge gathered in the past to know what to look for, e.g. to observe that hydrogen is given off by testing with a lighted splint (past knowledge of the effect together with hearing the result).

identification (*n*) (1) the action, carried out by experiment, of determining the properties (p.27), hence the name, of a substance. (2) the action of naming a process (p.129) or a form of energy by comparing its characteristics (p.147) with those of a known process or form of energy, e.g. (a) the identification of a metal as magnesium by investigating (↑) its properties; (b) the identification of a radiation as X-rays by investigating its characteristics. **identify** (*v*), **identifiable** (*adj*).

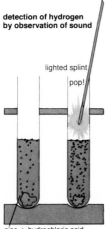

detection of hydrogen by observation of sound

lighted splint

pop!

zinc + hydrochloric acid

identical (*adj*) describes objects, substances, processes (p.129) and radiations (p.83) which have exactly the same number of properties (p.27) or characteristics (p.147) which are exactly the same, e.g. two crystals are identical if they both have the same colour, shape, and size, and are the same substance. **identity** (*n*).

similar (*adj*) describes objects, substances, processes (p.129) and radiations if they have many common properties or characteristics, but have a few different ones. **similarity** (*n*).

physical change a change in the state or in a physical property of a material while remaining the same material, e.g. (a) ice melting to water; (b) brass expanding on heating; (c) iron being magnetized.

chemical change a change in which new substances are formed with different properties, e.g. chalk, when heated, is chemically changed to lime and carbon dioxide.

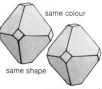

identical crystals

same colour

same shape

same size   same compound

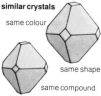

similar crystals

same colour

same shape

same compound

different size

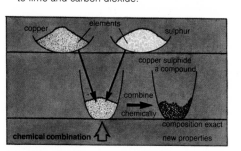

combination (*n*) the chemical union of two substances to form a new substance with different properties, e.g. the combination of iron and oxygen to form iron oxide (rust).
combine (*v*).

formation (*n*) (1) the action of bringing a material into being by a chemical change, e.g. the formation of ammonia gas on heating an ammonium salt with an alkali. (2) bringing any object into being by a physical change or effect, e.g. (a) the formation of dew by condensation; (b) the formation of an image by a lens. **form** (*v*).

composition (*n*) the elements (p.103) with their proportions (p.23) in a substance (↓) form its chemical composition. **be composed of** (*v*).

compound (*n*) a substance (↓) with a known composition of elements which cannot be separated by physical means, e.g. lime is a compound of calcium and oxygen (2 elements) in the proportion 40 parts calcium : 16 parts oxygen by mass. An exact formula (p.105) can be given for a compound.

substance (*n*) a material whose composition does not vary, but it may not be possible to know its exact formula (p.105), e.g. starch is a substance because its composition is known though its exact formula is not. Its formula is $(C_6H_{10}O_5)$ *n* where *n* is a large unknown number.

decompose (*v*) to break a substance or compound into simpler substances or compounds by chemical action, heat, or electric current, e.g. heat decomposes chalk into lime and carbon dioxide. **decomposition** (*n*).

mixture (*n*) a material made by mixing substances together; the substances can be in any proportion, and can be readily separated from each other by physical methods. Each substance keeps its own properties.

decomposition

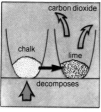

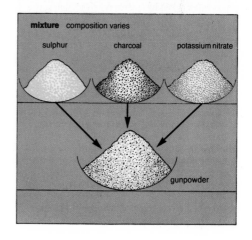

mixture   composition varies

sulphur   charcoal   potassium nitrate

gunpowder

constituent (*n*) a single substance or an element in a compound or mixture, e.g. (a) sulphur is a constituent of copper sulphide; (b) sulphur is a constituent of gunpowder. **constituent** (*adj*).

contain (*v*) to have as a constituent when some, but not all, the constituents are named, e.g. gunpowder contains sulphur and carbon.

consist of (*v*) to have as a constituent when all the constituents present in a mixture or compound are named, e.g. gunpowder consists of sulphur, charcoal and potassium nitrate.

chemical reaction a chemical change (p.94) that takes place when two or more substances are put together, e.g. the chemical reaction between zinc and sulphuric acid. **chemical reactivity, chemically reactive.**

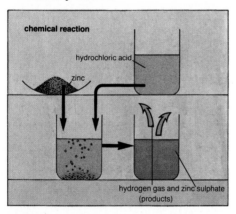

chemical reaction

hydrochloric acid

zinc

hydrogen gas and zinc sulphate (products)

product (*n*) a substance formed by a chemical reaction, e.g. hydrogen is a product of the reaction between zinc and sulphuric acid.

reversible reaction a chemical reaction in which the products can react chemically with each other to form the original substances, so the reaction can go in either direction, e.g. steam and iron react to form hydrogen and an oxide of iron; the oxide of iron and hydrogen react to form steam and iron.

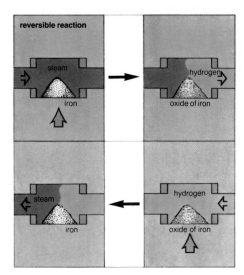

reversible reaction

steam
iron

hydrogen
oxide of iron

steam
iron

hydrogen
oxide of iron

chain reaction a chemical reaction which produces a product and the product causes a second reaction producing a second product which causes a third reaction and so on, e.g. in nuclear fission (p.109) an atom produces a product and two neutrons (p.106); each neutron reacts with another atom to produce a product and two neutrons, so the reaction continues faster and faster.

effervesce (*v*) to produce a large quantity of bubbles (p.92) of gas by a chemical reaction, e.g. when an acid acts on chalk the liquid effervesces. **effervescence** (*n*), **effervescent** (*adj*).

replaceable (*adj*) describes, in an acid, hydrogen atoms (p.103) which can be displaced and metal atoms put in their place, e.g. the hydrogen in sulphuric acid is replaceable by zinc.

dehydrate (*v*) to remove water, whether present as moisture or chemically combined, by heat or by a chemical reaction, e.g. to dehydrate crystals of copper (II) sulphate by heat or by the action of concentrated sulphuric acid.

effervescence

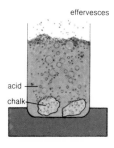

effervesces

acid
chalk

volatile (*adj*) of a liquid, evaporates readily, e.g. petrol is a volatile liquid. **volatility** (*n*).

odour (*v*) the property of a substance recognized by smell, e.g. the odour of petrol. **odorous** (*adj*), **odoriferous** (*adj*).

odourless (*adj*) describes a substance without odour.

deliquescent (*adj*) describes a solid which tends to absorb so much water vapour from the atmosphere that it forms a solution. **deliquescence** (*n*).

hygroscopic (*adj*) describes a solid which tends to absorb water vapour from the atmosphere and becomes wet, e.g. common salt left in humid air.

efflorescent (*adj*) describes crystals which tend to lose chemically combined water to the atmosphere, e.g. crystals of sodium carbonate are efflorescent and decompose to a white powder. **efflorescence** (*n*).

reagent (*n*) a substance which produces a chemical reaction with a certain chemical, and can be used in testing to discover whether that chemical is present, e.g. silver nitrate is a reagent for testing for the presence of a chloride.

agent (*n*) a substance or solution used for a particular chemical process, e.g. an oxidizing (↓) agent to oxidize iron (II) sulphate.

oxidation (*n*) (1) the addition of oxygen to a substance. (2) the removal of hydrogen from a substance. (3) the increasing of positive electrovalency (p.109), e.g. (a) the oxidation of copper to copper (II) oxide; (b) the oxidation of hydrogen chloride to chlorine; (c) the oxidation of iron (II) sulphate to iron (III) sulphate. **oxidize** (*v*).

reduction (*n*) (1) the removal of oxygen from a substance. (2) the addition of hydrogen to a substance. (3) the decreasing of positive electrovalency (p.109), e.g. (a) the reduction of copper (II) oxide to copper; (b) the reduction of chlorine to hydrogen chloride; (c) the reduction of iron (III) sulphate to iron (II) sulphate. **reduce** (*v*).

common reagents

silver nitrate | sulphuric acid

reagent bottle

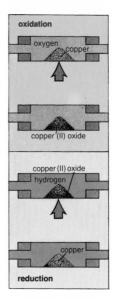

oxidation

oxygen | copper

copper (II) oxide

copper (II) oxide

hydrogen

copper

reduction

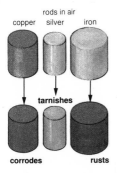

rods in air
copper   silver   iron

tarnishes

corrodes   **rusts**

bleaching

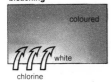

coloured

white

chlorine

corrosion (*n*) the slow destruction of a metal by chemical action such as an acid or atmospheric oxygen, e.g. copper corrodes in the air and forms a green coat. **corrode** (*v*).

corrosive (*adj*) describes an agent of corrosion of a metal, and also a substance which attacks animal tissues (p.140).

rust (*n*) the coat of oxide in iron formed by corrosion. **rust** (*v*).

tarnish (*v*) of bright, shining metals to corrode, e.g. silver tarnishes and becomes dull.

bleach (*v*) to make an object white by destroying its colour, e.g. chlorine bleaches cotton to make it white; chlorine is a bleaching agent.

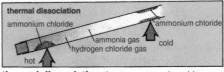

thermal dissociation
ammonium chloride · · · · · · ammonium chloride
ammonia gas   cold
hydrogen chloride gas
hot

thermal dissociation the temporary breaking down by heat of a substance into simpler substances; on cooling, these simpler substances combine to form the original substance, e.g. ammonium chloride under thermal dissociation forms ammonia and hydrogen chloride gases; on cooling, these two gases combine to form ammonium chloride.

pyrolysis (*n*) the decomposition by heat of a substance into simpler substances which do not combine on cooling. **pyrolitic** (*adj*).

catalysis (*n*) the increasing of the rate of a chemical reaction by a **catalyst**, a substance which is not itself changed, e.g. platinum increases the rate of sulphur dioxide combining with oxygen; catalysis has increased the rate of reaction. **catalyst** (*n*), **catalyze** (*v*), **catalytic** (*adj*).

violent (*adj*) describes a chemical reaction which is almost explosive in its rate. **violence** (*n*).

impurity (*n*) a small amount of a foreign substance in a large amount of another substance, e.g. lead is often an impurity in silver, i.e. silver contains a small amount of lead.

purify (*v*) to remove impurities.

electrolysis (*n*) the chemical decomposition
(p.95) of a substance by an electric current;
the substance is either in solution in water or is
molten (p.40). **electrolyze** (*v*), **electrolytic** (*adj*).

electrolyte (*n*) (1) any substance which, either
in solution or when molten, conducts (p.74)
an electric current and is decomposed by the
current. (2) a liquid in an electric cell or a
voltameter (↓). Soluble inorganic salts (p.115)
are electrolytes, e.g. sodium chloride.

non-electrolyte (*n*) any solid substance in
solution, or molten, or any liquid substance
which does not conduct an electric current,
e.g. sugar, petrol, alcohol are non-electrolytes.

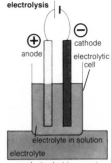

electrolysis

anode

cathode

electrolytic
cell

electrolyte in solution

electrolyte

− conducts electric current
− is decomposed by current

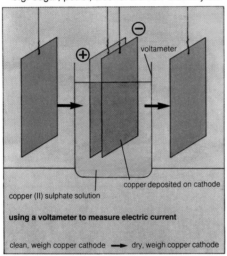

voltameter

copper deposited on cathode

copper (II) sulphate solution

using a voltameter to measure electric current

clean, weigh copper cathode ➡ dry, weigh copper cathode

voltameter (*n*) a vessel in which electrolysis
takes place for the purpose of measuring electric
current by weighing the mass of a metal
deposited (p.102) on a cathode (↓).

electrolytic cell any vessel used for electrolysis.

anode (*n*) the positive electrode (p.85) of an
electrolytic cell. **anodic** (*adj*).

cathode (*n*) the negative electrode of an
electrolytic cell. **cathodic** (*adj*).

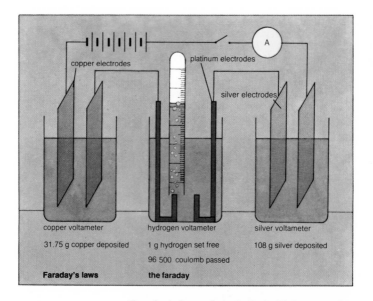

copper electrodes

platinum electrodes

silver electrodes

A

copper voltameter

31.75 g copper deposited

Faraday's laws

hydrogen voltameter

1 g hydrogen set free

96 500 coulomb passed

the faraday

silver voltameter

108 g silver deposited

Faraday's laws of electrolysis (1) the mass of a substance set free or deposited (p.102) is proportional to the strength of the electric current and to the time the current flows, i.e. to the quantity of electric charge passed. (2) the masses of different substances set free by the same quantity of electric charge (measured in coulomb) are proportional to their chemical equivalents (p.104), e.g. 96 500 coulomb of current will set free 1 g of hydrogen, or 8 g oxygen, or 108 g silver, or 12 g magnesium.

ion (n) an atom (p.103), or a group of chemically combined atoms, carrying an electric charge, either positive or negative. A positive ion is formed by an atom, or group of atoms, losing one or more electrons (p.106). A negative ion is formed by an atom, or group of atoms, gaining one or more electrons. Ions carry the electric current between the electrodes in electrolysis. Symbols: Cu^{2+}, Na^+, Cl^-, SO_4^{2-}. **ionic** (adj).

ionize (v) to form ions. **ionization** (n).

the ionic theory

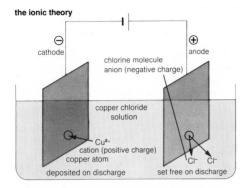

cathode ⊖   ⊕ anode

chlorine molecule
anion (negative charge)

copper chloride
solution

Cu²⁺
cation (positive charge)
copper atom
deposited on discharge

Cl⁻   Cl⁻
set free on discharge

anion (*n*) an ion with a negative charge; it is attracted to the anode (p.100) in electrolysis, and is discharged at the anode, e.g. chlorine ion, Cl^-.

cation (*n*) an ion with a positive charge; it is attracted to the cathode (p.100) in electrolysis, and is discharged at the cathode, e.g. copper ion, Cu^{2+}.

electrochemical equivalent the mass of a substance set free or deposited (↓) by 1 coulomb of electric charge, i.e. by a current of 1 ampere flowing for 1 second. Abbreviation: e.c.e.

deposit (*n*) (1) a layer of a metal put down on the cathode during electrolysis (p.100), e.g. a deposit of copper formed on the cathode in the electrolysis of copper (II) sulphate. (2) solid material left on a surface by a current of a fluid, e.g. (a) a deposit of mud at the mouth of a river; (b) a deposit of dust, from air, on surfaces. **deposit** (*v*), **deposition** (*n*).

faraday (*n*) a constant quantity of electric charge: approximately 96 500 coulombs. 1 faraday deposits 108 g silver or 31.7 g copper or sets free 1 g of hydrogen. It is equivalent to 1 mole (p.105) of electrons (p.106). Symbol: *F*.

electroplating (*n*) the deposition of a thin layer of a metal on another metal, for decoration or protection using an electrolytic process, e.g. the electroplating of steel with chromium.

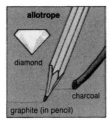

allotropes of carbon

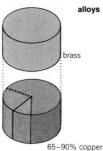

65–90% copper
35–10% zinc

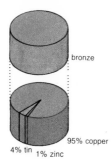

95% copper
4% tin 1% zinc

70% mercury
30% copper

dentist's amalgam

element (*n*) a substance which cannot be decomposed (p.95) by chemical reaction into simpler substances. It consists of atoms (↓) of the same atomic number (p.107). (See periodic table on front endpapers.)

allotrope (*n*) one of the different physical forms of an element, but possessing the same chemical properties as other allotropes, e.g. the allotropes of carbon include diamond, graphite and charcoal, all with the same chemical properties. **allotropic** (*adj*).

metal (*n*) an element (↑) whose atoms (↓) form positive ions and which generally has the properties of lustre (↓), ductility (↓) and malleability (↓), and of being a good conductor of heat and electric current. **metallic** (*adj*).

non-metal (*n*) an element not possessing the general properties of a metal. Many non-metals are gases, some are solids and one is a liquid, at room temperature. Some form negative ions.

ductile (*adj*) describes a substance that can be pulled into a wire while cold and under pressure. **ductility** (*n*).

malleable (*adj*) describes a substance that can be beaten or rolled into thin sheets. **malleability** (*n*).

lustre (*n*) the property of being bright enough to be used as a mirror, e.g. silver has a lustre. **lustrous** (*adj*).

atom (*n*) the smallest particle (p.26) of an element which has the properties of that element, and takes part in chemical reactions.

molecule² (*n*) the smallest particle of an element or a compound which can exist free and has the properties of that element or compound. A molecule of an element consists of one or more atoms; a molecule of a compound consists of one or more atoms of each element composing the compound, e.g. a molecule of hydrogen consists of two atoms of hydrogen; a molecule of water consists of two atoms of hydrogen and one atom of water. **molecular** (*adj*).

alloy (*n*) a material composed (p.95) of two or more metals, or composed of a metal and a non-metal, e.g. brass, steel, bronze are alloys. The composition of an alloy can vary slightly.

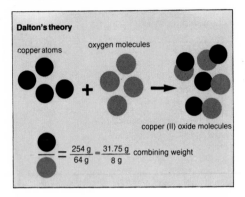

Dalton's theory the theory explains the laws of chemical combination (p.94) by saying that all substances are composed of atoms. The atoms of any one element are identical (p.94), but are different from the atoms of all other elements, at least in mass. Compounds are formed by the chemical combination of atoms in simple proportions. Present theories have different descriptions of atoms (p.103), but Dalton's theory is suitable for explaining the laws of chemical combination.

relative atomic mass the ratio of the average mass of an atom of an element to 1/12 of the mass of a carbon atom. Average mass is used because there may be isotopes (p.106) of the atoms.

relative molecular mass the ratio of the mass of a molecule of a substance to 1/12 of the mass of a carbon atom. The relative molecular mass of a compound is equal to the sum of the relative atomic masses of all the atoms in the molecule

combining weight the mass in grammes of an element that will combine with or replace 1 g of hydrogen or 8 g of oxygen.

chemical equivalent (1) the combining weight of an element. (2) of an acid, the mass of acid (p.114) containing 1 g of replaceable hydrogen; of a base (p.114), the mass that neutralizes (p.114) the chemical equivalent of an acid.

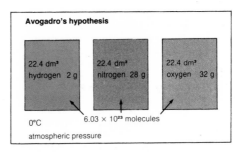

Avogadro's hypothesis

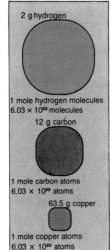

mole

Avogadro's hypothesis equal volumes of all gases, under the same conditions of temperature and pressure, contain the same number of molecules.

mole (*n*) the S.I. unit for amount of substance. 1 mole is the amount of substance which contains as many elementary units (i.e. particles) as there are atoms (p.103) of carbon in 0.012 kg of carbon-12. The elementary units must be named and can be molecules, atoms, ions (p.101), radicals (p.116), electrons (p.106) or other particles. Carbon-12 is an isotope (p.106) of carbon.

monatomic (*adj*) describes a molecule of an element which consists of one atom only, e.g. helium and neon are monatomic gases.

diatomic (*adj*) describes a molecule of an element which consists of two atoms, e.g. the molecules of hydrogen, oxygen, chlorine.

symbol (*n*) a letter or a small line diagram which represents a quantity (p.13), an instrument or device (p.10), a chemical element (p.103), or a process in mathematics, e.g. *p* stands for pressure;—⌐—stands for a switch; Fe stands for iron; ÷ stands for divide.

formula (*n*) chemical symbols written together to show the atoms in a molecule of a compound or in an ion, e.g. (a) the formula MgO stands for a molecule of magnesium oxide and shows it is composed of 1 atom of magnesium combined with 1 atom of oxygen; (b) the formula NO_3^- stands for a nitrate ion. The formula gives the composition of a substance.

electron (*n*) a very small particle, part of all atoms; it has the smallest possible negative electric charge (1.6×10^{-19} coulomb), and a mass approximately 1/1840 that of a hydrogen atom. An electron can be set free from an atom and travel by itself, e.g. as in cathode-ray tubes (p.86). **electronic** (*adj*).

proton (*n*) a very small particle, part of all atoms; it has the smallest possible positive charge, exactly equal to that of an electron; and a mass approximately 1839/1840 that of a hydrogen atom. A proton is a hydrogen ion.

neutron (*n*) a very small particle, part of all atoms except hydrogen; it has no electric charge; its mass is approximately equal to that of a proton.

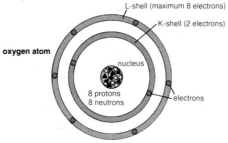

oxygen atom

L-shell (maximum 8 electrons)
K-shell (2 electrons)
nucleus
8 protons
8 neutrons
electrons

nucleus[1] (*n*) (*nuclei*) the central part of an atom, containing protons and neutrons, except for hydrogen which has a nucleus of one proton. The mass of an atom is almost all in its nucleus, as the electrons which surround the nucleus have a very much smaller mass. The nucleus has a positive charge (from its protons) which, in an atom, is exactly equal to the total charge of the electrons which surround it. **nuclear** (*adj*).

isotope (*n*) one, of two or more atoms of an element, which contains the same number of protons as the other atoms, i.e. has the same atomic number, and the same chemical properties, but has a different number of neutrons and hence a different atomic mass. **isotopic** (*adj*).

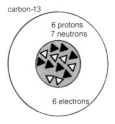

two isotopes of carbon
carbon-12

6 electrons

6 protons
6 neutrons

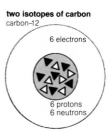

carbon-13

6 protons
7 neutrons

6 electrons

electron shell electrons surround the nucleus (↑), and are grouped together in different shells. Each shell is at a different distance from the nucleus, and is filled by a definite number of electrons. The shells are considered to be spheres surrounding the nucleus and the outer shell gives an atom its volume. The diameter of an outer shell is about 100 000 times greater than the diameter of the nucleus.

K-shell the shell of electrons nearest the nucleus; it is filled by 2 electrons.

L-shell the next shell, after the K-shell; it is filled by 8 electrons.

M-shell the next shell after the L-shell; it is filled by 18 electrons.

atomic number the number of protons in a nucleus (↑). It determines the chemical nature of the atom, i.e. it determines to which element an atom belongs. The symbol for atomic number is Z.

table

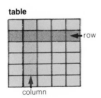

column

periodic system the elements, when arranged along rows in a table of increasing atomic number, form groups in the columns of the table; the elements in a group have similar chemical properties. This regular repeating of properties is the periodic system. (See periodic table on front endpapers.)

transition elements these elements are in the middle of a row in the periodic system. They use an inner shell to provide valency electrons (p.106). The elements are metals with more than one electrovalency (p.109).

radioactivity (*n*) the property of atomic nuclei (↑) of spontaneous (p.112) disintegration (↓). As the nuclei disintegrate, they emit (p.45) alpha or beta particles (p.108) or gamma rays (p.108). Radioactivity takes place mainly in elements of high atomic number. It is not altered by changes in pressure or temperature. **radioactive** (*adj*).

disintegrate (*v*) (1) to break into small pieces because of physical force being applied, e.g. rocks disintegrate because of the effects of wind and water. (2) of radioactive nuclei to emit one or more atomic particles, e.g. protons. **disintegration** (*n*).

range (*n*) (1) the distance to which an object or particle can travel, e.g. the range of a bullet from a gun. (2) the difference between the highest and the lowest of a set of values, e.g. the range of an ammeter. (3) the land area over which an organism is found.

alpha particle a positively charged particle consisting of two protons and two neutrons, identical with the nucleus of a helium atom. Its range in air is about 6 cm, and it is emitted (p.45) by some radioactive (p.107) nuclei. Alpha particles form alpha rays.

beta particle an electron with a high velocity emitted (p.45) from a radioactive nucleus; this happens when a neutron in the nucleus changes into a proton and an electron, and the electron is emitted as a beta particle. The range in air is about 750 cm. Beta particles form beta rays.

gamma rays radiation (p.83) with wavelengths shorter than those of X-rays. The rays are emitted (p.45) by radioactive nuclei together with either alpha or beta particles. There is no limit to their range in air. Their energy is decreased by half each time they pass through lead 1 cm thick.

Geiger counter an instrument which tests for radioactivity and measures the strength of alpha, beta or gamma rays, *see diagram*. Rays enter the window of the counter and discharge the potential between anode and cathode. The discharge is heard in a telephone receiver. The number of discharges is proportional to the strength of the rays.

effect of a magnet on radioactive radiation

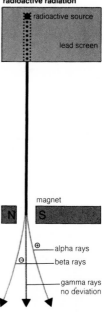

Alpha and beta rays are deflected by magnetic fields because they are electrically charged. Gamma rays are not deflected as they have no electric charge. The illustration shows the deviation of rays caused by a magnet.

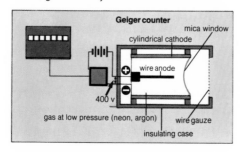

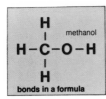

bonds in a formula

H
|
H–C–O–H   methanol
|
H

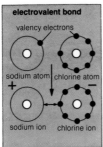

electrovalent bond

valency electrons

sodium atom chlorine atom

+ —

sodium ion   chlorine ion

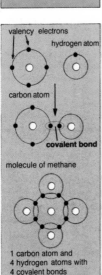

valency electrons

hydrogen atom

carbon atom

covalent bond

molecule of methane

1 carbon atom and
4 hydrogen atoms with
4 covalent bonds

half-life the time taken for half the atoms in a piece of radioactive material to disintegrate. Each radioactive element has its own half-life period and the values have a range (p.108) from one millionth of a second to more than one million years.

nuclear fission the breaking of the nucleus of an element with a high atomic number (e.g. uranium) into two approximately equal parts, with the emission (p.45) of neutrons (p.106) and the setting free of large quantities of energy. Fission can be spontaneous (p.112) or caused by the bombardment (p.83) of nuclei by neutrons.

nuclear fusion the forming of an atomic nucleus from two nuclei of low atomic number; large quantities of energy are set free. The reaction takes place only at very high temperatures.

bond (*n*) the force which holds atoms or ions together to make molecules or crystal structures. There are two main kinds of bonds, electrovalent (↓) and covalent (↓).

valency electron an electron which takes part in a bond. Except for the transitional elements (p.107), valency electrons are in the outermost electron shell.

electrovalency (*n*) (1) the joining together of two atoms by electrostatic (p.71) charges. One atom loses one or more electrons and becomes positively charged, the other atom gains one or more electrons and becomes negatively charged. The unlike charges attract and form an electrovalent bond. (2) the number of electrons that an atom, or group of atoms, loses or gains; the number of electrovalent bonds it can form. **electrovalent** (*adj*).

shared (*adj*) of an object, held by two or more other objects, e.g. an electron shared by two atoms.

covalency (*n*) (1) the joining together of two atoms by the sharing of a pair of electrons. Each atom gives one electron to form the pair. The electron pair forms a covalent bond. (2) the number of covalent bonds an atom can form.

crystal (*n*) a glass-like piece of a solid with a regular shape, e.g. crystals of sugar. A substance which forms crystals is called a crystalline solid; all its crystals have the same shape. Electrovalent (ionic) crystals are formed by electrolytes (p.101) from ions; covalent crystals are formed by non-electrolytes from molecules; metallic crystals are formed by metals from atoms. **crystallize** (*v*), **crystalline** (*adj*).

crystallization (*n*) the formation of crystals from a warm, saturated solution.

water of crystallization a definite molecular proportion of water, chemically combined with ions or molecules, in crystals formed from solutions in water, e.g. in copper (II) sulphate the proportion of compound to water of crystallization is 1:5; the formula of the compound is $CuSO_4, 5H_2O$.

hydrated (*adj*) describes crystals with water of crystallization (↑).

anhydrous (*adj*) describes crystals with no water of crystallization (↑) or amorphous (↓) solids.

amorphous (*adj*) without regular shape, not having a crystalline structure, e.g. an amorphous solid is like a powder, such as flour.

tetrahedral (*adj*) with a shape like a tetrahedron, which has four flat triangular faces, *see diagram*. A regular tetrahedron has three equal sides for each triangle.

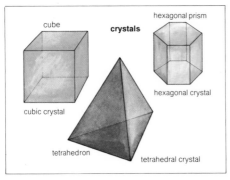

cube   **crystals**   hexagonal prism

hexagonal crystal

cubic crystal

tetrahedron   tetrahedral crystal

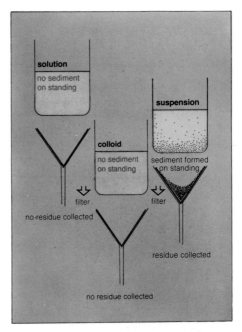

colloid (*n*) a substance that does not dissolve, nor is suspended (p.35) in a liquid, but is *dispersed* in a liquid. A colloid has particles larger than molecules or ions in solution, but smaller than those in suspensions (↓); the particles cannot be filtered (p.91) to form a residue, e.g. starch, glue, are colloids; they form **colloidal solutions**. **colloidal** (*adj*).

suspension (*n*) a liquid containing small solid particles suspended (p.35) in it. When filtered (p.91) the solid particles are collected as a residue; if left, the particles slowly **settle** to the bottom of the container and form a sediment, e.g. earth shaken up with water forms a suspension.

emulsion (*n*) a colloidal solution of one liquid in another liquid, e.g. oil shaken up with water forms an emulsion.

air (*n*) the mixture (p.95) of gases which surrounds the Earth and forms its atmosphere. Dry air consists of approximately 78% nitrogen, 21% oxygen, 0.03% carbon dioxide and the remainder the rare (also called noble or inert) gases such as helium, neon, argon.

combustion (*n*) a chemical reaction, between a substance and oxygen in the air, in which heat is given out, and a flame may be formed, i.e. burning. In *rapid combustion*, heat is given out at a high temperature, but in *slow combustion*, heat is given out at a lower temperature, and never with a flame. **combustible** (*adj*).

flame (*n*) a mass of gas so hot that it gives out light; the heat is produced by combustion.

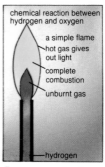

chemical reaction between hydrogen and oxygen

a simple flame

hot gas gives out light

complete combustion

unburnt gas

hydrogen

flame

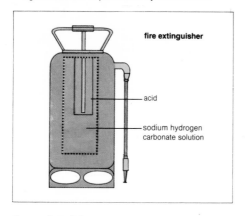

fire extinguisher

acid

sodium hydrogen carbonate solution

fire-extinguisher (*n*) a device that supplies a fluid which stops combustion, e.g. an extinguisher using carbon dioxide foam.

explosion (*n*) a sudden expansion (p.38) of gases, produced by rapid combustion, which exerts a very strong force when shut in a small space. **explode** (*v*), **explosive** (*adj*).

inflammable (*adj*) can burn easily and readily when a flame is applied.

spontaneous (*adj*) caused by itself, e.g. spontaneous combustion has no outside cause, a substance starts to burn due to causes inside itself.

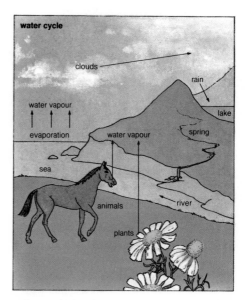

water cycle

clouds

rain

water vapour

lake

evaporation

water vapour

spring

sea

animals

river

plants

water-cycle the changes in the state of water:
(1) evaporation (p.42) from rivers, lakes and the
sea and respiration (p.191) of plants and animals
producing water vapour; (2) water vapour
condensing to form clouds; (3) rain falling from
clouds; (4) rain water passing through earth to
rivers and lakes and on to the sea. Each event
follows the one before it, so the changes form a
continuous process.

hard water water containing salts (p.115) of calcium,
magnesium or both metals. With soap, hard
water does not form a **lather** (lots of bubbles)
immediately; instead it forms a **scum** (particles
of solid on the water surface) first and a lather
later. Hard water which can be softened (↓) by
boiling is called temporary hard water; hard
water which can be softened by adding sodium
carbonate is called permanent hard water.

soft water water not containing salts of calcium
or magnesium. With soap, soft water forms a
lather (↑) immediately. **to soften water**.

acids and bases

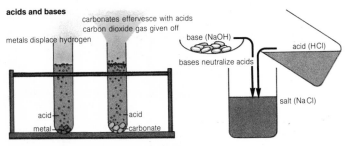

carbonates effervesce with acids
carbon dioxide gas given off

metals displace hydrogen

base (NaOH)

acid (HCl)

bases neutralize acids

salt (NaCl)

acid
metal
acid
carbonate

acid (*n*) a substance which (1) forms hydrogen
ions (p.101) in solution; (2) contains hydrogen
which can be replaced by a metal to form a
salt (↓). Some acids are corrosive (p.99) and
most acids change an indicator (p.116). **acidic**
(*adj*), **acidify** (*v*).

base (*n*) a substance which reacts with an acid to
form a salt and water only, generally an oxide or
a hydroxide of a metal. **basic** (*adj*).

alkali (*n*) a soluble base (↑) which forms
hydroxyl (↓) ions (p.101) in solution.

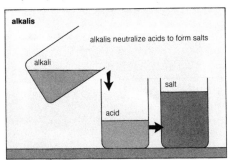

alkalis

alkalis neutralize acids to form salts

alkali

salt

acid

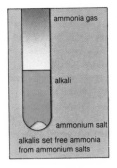

ammonia gas

alkali

ammonium salt

alkalis set free ammonia
from ammonium salts

hydroxyl (*n*) the group of atoms –OH; the
hydroxyl ion (OH$^-$) has a negative charge and an
electrovalency of 1.

neutralization (*n*) the reaction between an acid
and a base, in which both lose their properties,
and a salt is formed. **neutralize** (*v*), **neutral** (*adj*),
e.g. hydrochloric acid neutralizes sodium
hydroxide and sodium chloride is formed.

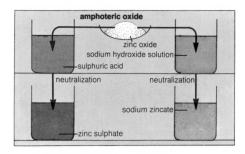

amphoteric oxide
zinc oxide
sodium hydroxide solution
sulphuric acid
neutralization · neutralization
sodium zincate
zinc sulphate

amphoteric (*adj*) describes a substance which can neutralize an acid and also neutralize a base, e.g. zinc oxide neutralizes acids to form zinc salts and it neutralizes bases to form zincates.

salt (*n*) a compound formed when the hydrogen of an acid is replaced by a metal. The salt is named from the metal and the acid; if soluble, it produces ions (p.101) in solution, a cation from the metal and an anion from the acid.

acid salt a salt formed when there is not enough acid to neutralize (↑) a base completely. Acid salts are formed only with dibasic (↓) acids, as only part of the hydrogen is replaced by a metal, e.g. sodium hydrogen sulphate.

basic salt a salt formed when there is not enough base to neutralize (↑) an acid completely; it consists of a neutral salt with a definite proportion of the base, e.g. basic lead carbonate $2PbCO_3$, $Pb(OH)_2$.

basicity (*n*) (1) the number of hydrogen atoms in a molecule of an acid which can be replaced by a metal, e.g. ethanoic (acetic) acid has the formula CH_3COOH but only one hydrogen atom can be replaced, so its basicity is 1. (2) the number of moles (p.105) of hydrogen ions formed from one mole of an acid, e.g. 1 mole of sulphuric acid forms 2 moles of hydrogen ions, its basicity is 2.

monobasic (*adj*) of an acid, having a basicity of 1, e.g. hydrochloric acid.

dibasic (*adj*) of an acid, having a basicity of 2, e.g. sulphuric acid. Dibasic acids can form acid salts.

indicator (*n*) a substance that changes colour when put in acids and alkalis; it is used to determine when neutralization has been completed, e.g. litmus (↓). **indicate** (*v*).

litmus (*n*) a soluble purple substance obtained from plants. In acids it turns red and in alkalis it turns blue. In neutralization, litmus is usually put in some alkali (it turns blue) and acid added until the litmus just turns red, showing neutralization is complete.

pH value a value on a numerical scale from 0–14 indicating the concentration of hydrogen ions (p.101) and hydroxyl (p.114) ions. At a value of 7, the concentrations of both ions are equal and the solution is neutral. Between 0 and 7, a liquid is acidic, with the hydrogen ion concentration increasing as the pH value falls from 7 to 0. Between 7 and 14, a liquid is alkaline, with the hydroxyl ion concentration increasing as the pH value increases.

pH scale

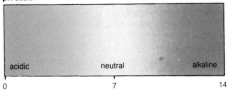

| acidic | neutral | alkaline |

| 0 | 7 | 14 |

radical (*n*) a group of atoms which can take part, without change, in chemical reactions. It is present in compounds, or as an ion in solution, but it is not found free. An acid radical and displaceable hydrogen form a molecule of an acid. In an ionized acid, the radical has one or more negative charges, e.g. (a) the sulphate radical, SO_4, in calcium sulphate, $CaSO_4$; (b) the sulphate ion, SO_4^{2-}, present in solution.

inorganic (*adj*) from mineral (p.119) sources. Inorganic substances and materials include all elements and all their compounds except for compounds containing carbon; carbonates, however, are considered as inorganic compounds.

naming compounds

prefix	number	example	
mono-	1	carbon monoxide	CO
di-	2	carbon dioxide	CO_2
		sulphur dioxide	SO_2
tri-	3	phosphorus trichloride	PCl_3
		phosphorus trioxide	P_2O_3
tetra-	4	tetrachloromethane	CCl_4
penta-	5	phosphorus pentachloride	PCl_5
hexa-	6	potassium hexacyanoferrate (III)	$K_3Fe(CN)_6$

oxide (*n*) a compound consisting of one element combined with oxygen, e.g. calcium oxide, CaO; nitrogen oxide, NO; iron (III) oxide, Fe_2O_3. The prefixes: mono-(one); di-(two); tri-(three); penta-(five) are used to show the covalency (p.109) of a non-metal, e.g. phosphorus pentoxide, P_2O_5.

peroxide (*n*) an oxide which forms hydrogen peroxide, formula H_2O_2, with a dilute acid instead of forming a salt and water, e.g. barium peroxide, BaO_2; the normal oxide is BaO.

hydroxide (*n*) a compound containing the group of atoms –OH; these compounds are bases, and, if soluble, form hydroxyl (p.114) ions in solution. When an alkali neutralizes an acid, the hydroxyl ions combine with hydrogen ions to form water. Examples are: sodium hydroxide, NaOH; calcium hydroxide, $Ca(OH)_2$.

chloride (*n*) the radical (↑) of hydrochloric acid (HCl). Chlorides are also considered as compounds consisting of one element combined with chlorine, e.g. phosphorus trichloride, PCl_3, sodium chloride, NaCl.

acid radicals and ions

radical	formula	ion
chloride	—Cl	Cl^-
bromide	—Br	Br^-
sulphate	—SO_4	SO_4^{2-}
sulphite	—SO_3	SO_3^{2-}
nitrate	—NO_3	NO_3^-
nitrite	—NO_2	NO_2^-
carbonate	—CO_3	CO_3^{2-}
sulphide	—S	S^{2-}
silicate	SiO_3	SiO_3^{2-}

sulphate (*n*) the radical (p.116) of sulphuric acid
(H_2SO_4); any salt formed from a metal and
sulphuric acid, e.g. sodium sulphate, Na_2SO_4.

sulphite (*n*) the radical (p.116) of sulphurous acid
(H_2SO_3); any salt formed from a metal and
sulphurous acid, e.g. sodium sulphite, Na_2SO_3.

nitrate (*n*) the radical (p.116) of nitric acid (HNO_3);
any salt formed from a metal and nitric acid,
e.g. sodium nitrate, $NaNO_3$.

nitrite (*n*) the radical (↑) having the group of
atoms —NO_2; the acid, formula HNO_2, cannot be
prepared. Salts include sodium nitrite, $NaNO_2$,
potassium nitrite, KNO_2.

carbonate (*n*) the radical (↑) of carbonic acid
(H_2CO_3); any salt formed by a metal and carbon
dioxide in solution, e.g. sodium carbonate,
Na_2CO_3.

rock (*n*) the solid material which forms the surface of the Earth. In some places, rock is covered by soil (p.230). There are different kinds of rock and each kind contains different minerals (↓). Examples of rocks are: granite (↓), chalk (p.128), coal (p.128), limestone (p.128). Rock can vary from very hard to soft; each kind of rock has its own particular hardness.

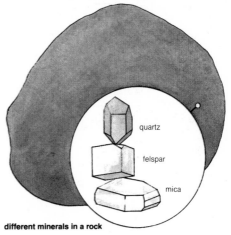

different minerals in a rock

mineral (*n*) a substance found in the ground; it has a particular chemical composition (p.95); it possesses chemical and physical properties by which it can be recognized. Examples of minerals are: (a) native elements such as gold and silver; (b) rock salt, which is sodium chloride; (c) iron pyrites, a kind of iron sulphide; (d) quartz (p.120); (e) mica, a silicate of aluminium, potassium, magnesium and iron. **mineral** (*adj*).

ore (*n*) a mineral (↑) dug from the earth, from which a useful substance, generally a metal, is obtained, e.g. haematite is an ore, and iron is obtained from it. Some ores are used to obtain non-metals and their compounds.

granite (*n*) a very hard igneous (p.121) rock containing at least three minerals: quartz (p.120), felspar, and mica.

basalt (*n*) a dark-coloured or black glass-like igneous (↓) rock formed by volcanic (p.125) action. It is the commonest type of lava, and contains silicates of iron, magnesium and calcium, and oxides of lime.

quartz (*n*) a mineral with clear, uncoloured crystals in the shape of hexagonal prisms (p.110), consisting of silicon dioxide. The most common substance in rocks.

silica (*n*) a very hard white substance, it is silicon dioxide. Quartz and sand consist of silica. Silica also combines with basic substances to form silicates, compounds which are common in minerals.

quartz crystal
a crystalline mineral

weathering (*n*) the action of wind, rain, ice, water, frost, or chemical substances on rock. Weathering loosens and breaks up the surface of the rock, setting free small pieces. These pieces can be carried by wind or rivers to other places. **weathered** (*adj*).

erosion (*n*) the destruction of the Earth's surface by strong weathering agents followed by the removal of the products of weathering. In particular, it is the removal of soil (p.230), leaving rocks uncovered, so that no plants grow. **erode** (*v*).

leaching (*n*) (1) the washing away of soluble mineral salts from soil (p.230), by water, usually rain water, passing quickly through the soil. Leaching produces a poor soil, with few plants, so erosion follows quickly. (2) washing out a soluble constituent (p.96) from a mixture. **leach** (*v*).

weathering

detritus (*n*) loose solids caused by weathering, which are carried by wind or water. Detritus usually consists of gravel, sand and clay. When it is no longer carried by wind or water, it settles and forms a sediment. **detrital** (*adj*).

silt (*n*) a material formed from very small pieces of rock; it is like mud. Particles of different sizes are given different names, *see diagram*. The particles of silt are larger than those in clay (p.230), but smaller than those in sand (p.230).

sediment (*n*) solids which separate from water and are deposited by gravity.

particle size	name	diameter
	boulder	>200 mm
	cobble	200–50 mm
	pebble	50–10 mm
	gravel	10–2 mm
	sand	2–0.1 mm
	silt	0.1–0.01 mm
	clay, dust	<0.01 mm

sedimentary rock any rock formed from the settling of detritus (sediment) or from sediment formed in layers by chemical action or from the changing of plant and animal bodies into decayed (p.146) material. For example: (a) sand, brought by a river, settles, and in time changes into hard sandstone; (b) chalk (p.128) is a soft rock formed from seashells which settled on the seabed; (c) trees from forests on the Earth 200 million years ago, decayed and changed into coal. Sedimentary rocks are usually soft and easily weathered. **sediment** (*n*), **sedimentary** (*adj*).

metamorphic rock a rock formed from sedimentary rock by pressure and high temperature. Great movements of the Earth break up sedimentary rock and push it down into the Earth, where it is very hot. In the Earth, the pressure and heat form hard, crystalline, metamorphic rock, from the sedimentary rock, e.g. marble is formed from limestone by heat and pressure. Metamorphic rocks are the most common in Earth's crust (p.124). **metamorphism** (*n*), **metamorphic** (*adj*).

igneous rock

basalt

igneous rock a rock which has solidified from liquid magma (p.124). The magma comes near to the surface (a) through volcanoes (p.125), forming volcanic rocks or (b) from being forced between metamorphic rocks, forming plutonic rocks. Basalt (↑) is the commonest volcanic rock and granite (p.119) the commonest plutonic (p.127) rock.

the rock cycle

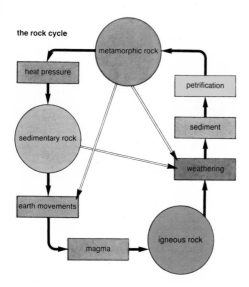

rock cycle the changing of types of rock into other types. The changes form a continuous process, *see diagram*.

petrification (*n*) the changing of plant and animal bodies, and sediment into rock, brought about by pressure and increased temperature. **petrify** (*v*).

stalactite (*n*) a finger-like shape of crystalline calcium carbonate hanging down from the roof of a cave in limestone rock. Water, containing dissolved calcium hydrogen carbonate, passes through the limestone and drops evaporate on the stalactite, increasing its length with a deposit (p.102) of calcium carbonate.

stalagmite (*n*) a finger-like shape of crystalline calcium carbonate pointing upwards from the floor of a cave in limestone rock. Drops from a stalactite (↑) fall on the stalagmite and evaporate, increasing its length. A stalagmite is below a stalactite and they both grow, eventually to meet.

formation of stalactites and stalagmites

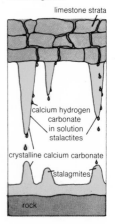

limestone strata

calcium hydrogen carbonate in solution

stalactites

crystalline calcium carbonate

stalagmites

rock

structure of Earth

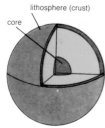

lithosphere (crust)

core

lithosphere (*n*) the outer, solid part of the Earth, which covers the hotter, inner parts. It is about 70 km thick. Metamorphic rocks (p.121) are the most common in the lithosphere, while sedimentary rocks are most common on its surface. The lithosphere is also known as the Earth's crust.

hydrosphere (*n*) the water portion of the Earth's crust as distinguished from the lithosphere, together with all the water vapour (p.41) in the atmosphere (p.51) around the Earth. The hydrosphere and the lithosphere together form the Earth's surface.

continent (*n*) a large, unbroken area of land. The lithosphere (↑) consists of the six great continental areas. **continental** (*adj*).

ocean (*n*) a large area of salt water, part of the hydrosphere (↑). The hydrosphere on the Earth's surface consists of four oceans. **oceanic** (*adj*).

sima (*n*) the lower layer of the lithosphere; it covers the whole of the Earth. The thickness of sima is not easily measured, *see diagram*. It has a density of 3000 kg/m^3 and is mainly made up of basalt (p.120). **simatic** (*adj*).

sial (*n*) the part of the lithosphere which stands on sima (↑). Sial is the base on which the land surfaces of the Earth stand; it is not continuous, as different parts of the sial are separated by the oceans. The thickness varies from very thick below high mountains to thin below plains. Sial has a density of about 2700 kg/m^3, and being less dense than sima, it floats on the sima. *See diagram*. **sialic** (*adj*).

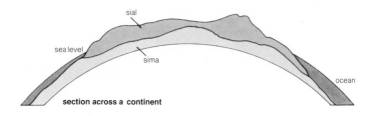

sial

sea level

sima

ocean

section across a continent

magma (*n*) a hot, sticky liquid state of rock made by conditions of high temperature and great pressure. Rocks from the lithosphere are forced down below the lithosphere (p. 123) by earthquakes (p. 126) and other earth movements and are changed into magma by the conditions there. Magma has a temperature of about 1000°C, and contains floating crystals and pockets of gas. Volcanic action and plutonic (p. 127) movements push magma up into the lithosphere. Magma cools to form igneous rocks (p.121).

core² (*n*) the innermost part of the Earth with a radius of about 3500 km; it has a density of between 8000 and 10 000 kg/m³ and is much denser than the lithosphere. The core is a mixture of iron and nickel with an inner, solid core, and an outer, liquid core.

crust (*n*) (1) a hard, outer cover over an inner softer core. (2) the lithosphere (p.123) of the Earth.

stratum (*n*) (*strata*) (1) one of a number of layers, each on top of the other. (2) an approximately level, or formerly level, layer of sedimentary rock. **stratification** (*n*).

fault (*n*) a break in the stratification of rocks; large earth movements cause the break, *see diagram*, and the strata (↑) are displaced, so that they no longer form a continuous layer.

seam (*n*) a stratum (↑) of an ore, or other useful mineral, e.g. a seam of coal.

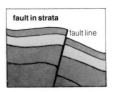

fault in strata

fault line

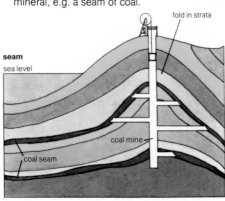

fold in strata

seam

sea level

coal mine

coal seam

volcano (*n*) a cone-shaped mountain formed when steam, lava (↓), gases, and rocks are forced out from inside the Earth by the pressure of gases and steam, usually with explosive force. **volcanic** (*adj*).

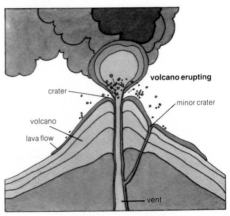

volcano erupting

crater

minor crater

volcano

lava flow

vent

crater (*n*) the hollow at the top of a volcano (↑). Minor craters may be formed on the sides of a volcano.

vent (*n*) (1) an opening or pipe to allow fluids to pass out. (2) the tube-like hollow in the middle of a volcano (↑), up which pass the steam, gases, lava and solid particles. When the volcano is no longer active, the vent is blocked by a *plug* of basalt (p.120).

lava (*n*) liquid magma (↑), usually white-hot, forced out of the vent (↑) of a volcano; and the solid magma formed when it cools.

pumice (*n*) a sponge-like (p.149), grey, glass-like solid formed as a crust on the top of lava when it cools. Many bubbles of gas in the lava form the sponge-like structure of pumice.

eruption (*n*) the explosive action of a volcano when gases, steam, and lava are thrown out of the vent (↑). **erupt** (*v*).

volcanic dust a black cloud of very small pieces of solid lava blown out during an eruption (↑).

earthquake (*n*) the shaking of the Earth's crust caused mostly by displacement along a fault (p. 124), or by volcanic action. Earthquakes only occur where such movements take place. They produce shock waves (↓) felt as vibrations of the Earth's surface and may destroy buildings. The place of maximum displacement is the focus of the earthquake. Of the many thousand earthquakes each year, only a few are destructive or even noticed by man.

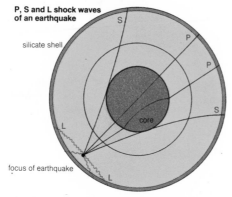

P, S and L shock waves of an earthquake

silicate shell

core

S
P
P
S
L
L

focus of earthquake

shock wave a wave motion (p.65) of compression or deformation (p.27) transmitted by rock structures in the Earth. The greater the rigidity (p.28) of a rock, the faster the waves travel through it. There are three kinds of shock waves, called P, S, and L waves. P waves are the fastest, travelling at approximately 650 km/hour; they pass through the Earth's core and are refracted by the different structural layers of the Earth, *see diagram*. S waves, slightly slower, follow the P waves, but do not pass through the Earth's core. L waves travel more slowly, round the Earth's surface, and have a much greater amplitude than the other two waves. The focus of an earthquake, from which shock waves start, is usually only a few kilometres below the Earth's surface. The point directly above the focus is the epicentre.

tremor (*n*) a vibration of the Earth's surface, caused by shock waves; a minor earthquake. Tremors are caused by P and S shock waves (↑) and are also called preliminary tremors. **tremble** (*v*).

secondary tremor a vibration of the Earth's surface caused by shock waves of large amplitude (L waves). These tremors cause almost all the damage from earthquakes.

seismograph (*n*) an instrument for recording tremors (↑); it produces a **seismogram** showing the amplitude of the tremors, *see diagram*.

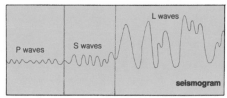

P waves   S waves   L waves

seismogram

tectonic (*adj*) formed in the Earth's crust, e.g. earthquakes are tectonic movements.

plutonic (*adj*) formed, or coming from, deep inside the Earth, below the Earth's crust, e.g. movements of magma are plutonic.

tide (*n*) the regular rise and fall of water in the sea or ocean, taking place twice in a lunar (p.62) day; it is caused by the attraction of the moon for the water, *see diagram*. A *high tide* is produced at a place opposite the moon, and also at a place on the far side of the Earth. A *low tide* is produced between the places of high tides on the Earth's surface. **tidal** (*adj*).

tides

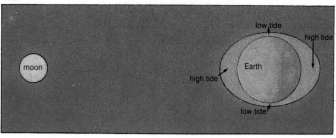

moon   low tide   high tide   Earth   high tide   low tide

chalk (*n*) a soft, white or grey variety of limestone
(↓) formed from the shells of very small sea
animals deposited on the sea bed. Chalk is
mainly used industrially (↓) to produce lime
(calcium oxide), the cheapest available alkali.
chalky (*adj*).

limestone (*n*) a medium hard, grey, sedimentary
(p. 121) rock, of calcium carbonate, formed from
sediment on shallow sea beds. Limestone is
used for buildings, for making cement and glass
and also for making lime.

coal (*n*) a soft, black, sedimentary (p.121) rock
formed by the petrification (p.122) of plant
material. The plants grew in the Carboniferous
period, over 200 million years ago. Coal consists
of carbon and other compounds; it burns slowly,
producing heat. Its chemical products are
obtained by destructive distillation (p.135).

petroleum (*n*) a liquid mineral containing many
different hydrocarbons (p.131); formed in
pockets in folds of rock strata (p.124); usually
found in layers of porous rock. About 260 litres
occur in a cubic metre of rock. It is thought to
have been formed from decaying plants and
animals, in a way similar to coal. The composition
varies from place to place. Petroleum is usually
under pressure from natural gas (↓) and when a
pipe is forced through the earth into a pocket of
petroleum, the pressure forces the petroleum up
the pipe. The chemical products of petroleum
are obtained by fractional distillation (p.134).
They include petrol, kerosene, lubricating oils
and asphalt.

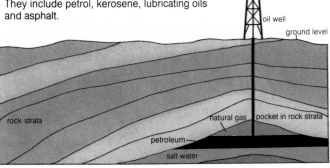

obtaining petroleum

oil well

ground level

rock strata

natural gas

pocket in rock strata

petroleum

salt water

natural gas gas formed in pockets above petroleum (↑); also gas sent out through volcanoes. In some places it consists mainly of methane (formula CH_4), in other places it consists of mixed hydrocarbons (p.131). It is used as a fuel (↓).

fuel (*n*) a material which can be burned to give out heat or to provide chemical energy. Most fuels are carbon compounds, e.g. wood, coal, petroleum.

raw materials materials obtained from natural sources (p.52) for use in making chemical substances, or other materials needed in industry (↓), e.g. (a) ores of iron are the raw material for making iron; (b) sea water is a raw material for making common salt; (c) sugar-cane is a raw material for making sugar.

industry (*n*) the making of materials, substances, articles in large amounts for selling to people, e.g. the clothing industry, the chemical industry, the steel industry. **industrial** (*adj*).

process¹ (*n*) (1) a set of events, following one after the other, all concerned with the same activity, e.g. the process of lighting a fire. (2) in chemistry, an industrial (↑) method of making a desired material or substance generally involving many operations (↓), e.g. (a) the Bessemer process for making steel; (b) the contact process for making sulphuric acid.

operate (*v*) to make a process (↑) or a machine work, e.g. (a) to operate the contact process so that sulphuric acid is produced; (b) to operate a hydraulic press. Simple machines, e.g. a lever, and simple processes, e.g. lighting a fire, are used, not operated. **operation** (*n*), **operator** (*n*), **operative** (*adj*).

by-product (*n*) a product other than the product for which a process (↑) is intended, e.g. a process intended to make sodium hydroxide also produces chlorine; the sodium hydroxide is the main product, while chlorine is the by-product.

waste product a substance made in a process (↑) but which has no industrial use.

contact process

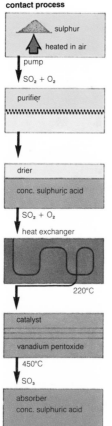

sulphur
heated in air
pump
$SO_2 + O_2$
purifier

drier
conc. sulphuric acid
$SO_2 + O_2$
heat exchanger
220°C

catalyst
vanadium pentoxide
450°C
SO_3
absorber
conc. sulphuric acid

smelt (*v*) to heat the ore of a metal with a suitable substance, usually coke, in order to obtain the metal from its ore, e.g. to smelt iron ore in a blast furnace (↓) to obtain iron. **smelting** (*n*).

furnace (*n*) a device to obtain strong heating effects using a fuel or electric current, e.g.
(a) a furnace to heat a boiler to produce steam;
(b) a furnace to smelt metals.

blast furnace a furnace which uses a forced current of air (a blast) to smelt ores of iron.

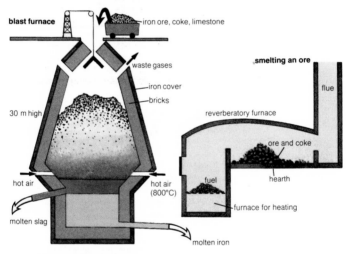

reverberatory furnace a furnace in which a material is heated by flames passing above it, and being directed down on to the material by the roof of the furnace, *see diagram*. It is used for smelting ores of metals.

slag (*n*) waste material, mainly non-metallic, formed during the smelting of ores, and floating on the molten metal. It contains silicates and other mineral material.

refine (*v*) to make as pure as possible, e.g.
(a) sugar is refined to remove unwanted plant material; (b) metals are refined to remove impurities. **refinement** (*n*), **refinery** (*n*).

carbon (*n*) an element, atomic number 6, relative atomic mass 12.01; it is a non-metal with three allotropic (p.103) forms. It has a covalency of 4 and the ability to combine its atoms in long chains and in rings; this allows a great variety of compounds to be formed, many of which are found in plants and animals.

organic (*adj*) describes materials, substances and compounds which contain carbon, other than the oxides of carbon and the carbonates. All organisms (p.147) consist of organic compounds.

hydrocarbon (*n*) a compound consisting of carbon and hydrogen only. The simplest hydrocarbon is methane, formula CH_4 (p. 109). Hydrocarbons can consist of long chains of carbon atoms, with hydrogen combined by covalent bonds (p.109), or they can consist of rings of carbon atoms with hydrogen combined, e.g. benzene (↓).

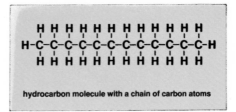

hydrocarbon molecule with a chain of carbon atoms

benzene ring the structure of a molecule of benzene in which six carbon atoms are joined by covalent bonds in the shape of a ring. The benzene ring is also seen in other compounds, e.g. in naphthalene and anthracene.

molecule of benzene

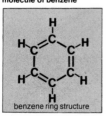

benzene ring structure

symbol for a benzene ring

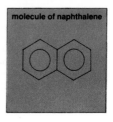

molecule of naphthalene

isomer (*n*) one of two compounds with the same number of atoms of each element contained in the molecule, but with different physical or chemical properties, e.g. ethanol and dimethyl ether both possess a molecular formula of C_2H_6O, but their structure differs, *see diagram*, and their physical and chemical properties are different. **isomeric** (*adj*), **isomerism** (*n*).

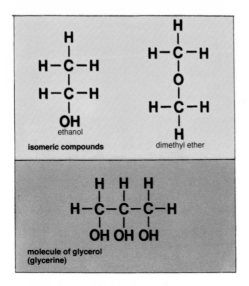

ethanol

isomeric compounds

dimethyl ether

molecule of glycerol (glycerine)

alcohol (*n*) a substance with the structure of a hydrocarbon but with one, or more, hydrogen atoms replaced by hydroxyl groups (p.114), e.g. ethanol is an alcohol, glycerol is an alcohol, *see diagram*. **alcoholic** (*adj*).

ester (*n*) a compound formed from an alcohol and an acid, usually an organic acid. The chemical reaction is: alcohol + acid → ester + water, e.g. ethanol and ethanoic (acetic) acid react to form ethyl ethanoate. **esterification** (*n*).

soap (*n*) an ester of glycerol and organic acids, mainly stearic, oleic, and palmitic acids. Soap is made by the action of alkalis on fats (p. 175).

polymer (*n*) a molecule with a high relative molecular mass (p. 104) formed by the chemical linking of many simpler molecules, each of the same substance, called monomers, e.g. ethene (ethylene) is a simple molecule, formula $CH_2:CH_2$; many molecules of ethene combine together to form polyethylene, formula: . . . $CH_2 \cdot CH_2 \cdot CH_2 \cdot CH_2$. . .; polyethylene is a polymer of ethene known as polythene.
polymerize (*v*), **polymeric** (*adj*).

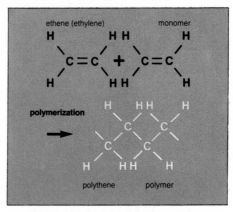

polymerization (*n*) the forming of a polymer from a simpler compound called a monomer.
polymerized (*adj*).
thermoplastic (*n*) a polymerized (↑) substance which becomes plastic (p.28) on heating and can be shaped by pressure and heat, without changing its chemical properties; the process can be repeated on further heating, e.g. polythene is a thermoplastic.
thermosetting (*adj*) describes a polymerized (↑) substance which first becomes plastic (p.28) on heating and then becomes hard because of a chemical change. A different substance is formed which cannot be made plastic by reheating. Thermoplastic and thermosetting substances are called **plastics**. They are shaped by heat and pressure, and keep their shape when cooled.

oil (*n*) any one of a number of greasy, combustible substances obtained from plants, animals or minerals. Oils and fats (p.175) from living organisms are esters (p.132) of glycerol and organic acids; Oils are liquid, and fats are solid, at room temperature. Mineral oil consists of hydrocarbons and is obtained from petroleum (p. 128). **oily** (*adj*).

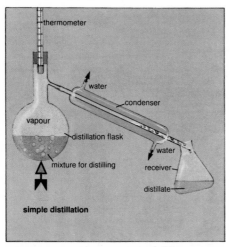

simple distillation

distillation (*n*) a physical process for separating (p.91) volatile (p.98) liquids from mixtures. The mixture is heated, the liquid is vaporized (p. 41); the vapour is then condensed (p.42) and collected. Liquids with different boiling points can be separated by the process. **distil** (*v*), **distillate** (*n*).

fractional distillation a distillation (↑) process for separating liquids with boiling points close together. A **fractionating column** is used, and vapour is taken out at different levels of the vertical column; the vapour is condensed and the different liquids are *fractionated* by the process. Fractionating is used to separate (p. 91) the different substances in petroleum (p.128).

fractionate (*v*) to split into fractions.

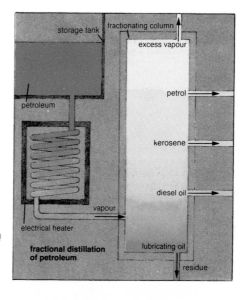

fractional distillation of petroleum

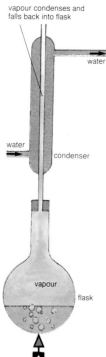

refluxing a mixture of organic liquids

vapour condenses and falls back into flask

water

water

condenser

vapour

flask

destructive distillation the distillation (↑) of volatile (p.98) substances from a solid, which is decomposed by the process, e.g. coal is destructively distilled leaving a solid residue, **coke**, while **coal-tar** and **pitch** are collected as a distillate, and coal-gas is collected in a gasometer.

still (*n*) an apparatus for distillation, especially for distilling alcohol or petroleum; also the vessel in which a mixture is heated for distillation.

reflux (*v*) to boil a liquid with a condenser above the liquid so that the vapour is condensed and falls back into the boiling liquid. Refluxing is a process for making chemical reactions of volatile (p.98) organic compounds take place. **reflux** (*adj*).

cracking (*n*) a kind of pyrolysis (p.99) used with mineral oils (↑) of high boiling point. The oils are passed through a red hot tube, decomposition takes place and mineral oils of lower boiling point, e.g. petrol, are produced.

fermentation (*n*) a chemical change caused by yeasts (p.146) and bacteria (p.145), in which a carbohydrate (p.173), usually a sugar, is changed to ethanol (ethyl alcohol) and carbon dioxide. Enzymes (p.167) catalyse the change. **ferment** (*v*).

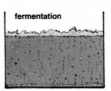

fermentation

saponification (*n*) the hydrolysis (↓) of an ester (p.132) using an alkali. An alcohol and a salt of the organic acid, or acids present, are formed. **saponify** (*v*).

hydrolysis (*n*) the chemical decomposition (p.95) of a substance by water, e.g. (a) some salts, such as iron (III) chloride, are partly hydrolyzed by water to form an acidic, and not a neutral, solution; (b) esters are hydrolyzed by water into an alcohol and an acid; the chemical reaction is slow, so an acid or an alkali is added to make the reaction go faster. **hydrolyze** (*v*), **hydrolytic** (*adj*).

substitution (*n*) the replacing (p.97) of an atom in a molecule of an organic compound by an atom of a different element or by a group of atoms, e.g. to replace a hydrogen atom in methane (CH_4) by a chlorine atom, forming chloromethane (CH_3Cl). **substitute** (*v*).

vulcanization (*n*) the process of making natural rubber harder and less elastic, so that it keeps its shape when in use. Rubber is heated with sulphur to form vulcanized rubber, as used in tyres. **vulcanize** (*adj*).

synthesis (*n*) the combining (p.94) of elements to form a simple compound, or of simple compounds to form a compound with many atoms in its molecules, e.g. the synthesis of explosives (p.112) such as *dynamite*, which is made from glycerol and nitric acid. **synthesize** (*v*).

synthetic (*adj*) made by chemical processes from simpler compounds, not obtained from natural products, e.g. synthetic threads, such as nylon (↓) and artificial silk, used in making clothes.

nylon (*n*) a polymer (p.133) with a long chain of carbon atoms to which *amide groups* ($-CONH_2$) are combined at intervals. There are many different polymers, with different physical properties, all called nylon. A common kind is the nylon used for thread.

Biology

cell (*n*) the smallest part of a plant or animal; some organisms (p.147) consist of one cell only, e.g. amoebae and bacteria, while others, e.g. trees, humans, contain many millions of cells. A cell consists of protoplasm (↓) with a membrane (↓) around it. It has the ability to take in chemical substances and use them to make various substances it needs for the process of living. Many cells have the ability to divide and form two new cells. **cellular** (*adj*).

protoplasm (*n*) a grey, jelly-like material containing many organic (p.131) chemical compounds, and organelles (p.141). Protoplasm is divided into cytoplasm (↓) and a nucleus (↓). **protoplasmic** (*adj*).

membrane (*n*) (1) a thin skin-like piece of material, covering or supporting a part of a plant or animal. (2) a cell-membrane is a very thin membrane (about 10 nm thick) composed of fat (p.175) and protein (p.172); it covers a cell and allows some substances to pass through into the cell, but not others. If the cell-membrane is damaged, the cell is destroyed. **membranous** (*adj*).

cell wall a wall of cellulose (↓) round the cell of a plant. It is formed by the protoplasm and is tight against the cell-membrane, *see diagram*. The cell wall is rigid (p.28) and gives a plant its strength to stand upright.

cellulose (*n*) a long carbon-chain polymer (p.133) of glucose (p.174). Cellulose molecules bind together to form a strong structure.

cytoplasm (*n*) all the protoplasm (↑) in a cell, except the nucleus (↓). It contains many different organic (p.131) chemical compounds and organelles (p.141). It is always in a state of movement, with a continuous change of compounds taking place. It is divided into ectoplasm and endoplasm. **cytoplasmic** (*adj*).

ectoplasm (*n*) (1) in plants, another name for cell-membrane. (2) in animal cells, it is the outer layer of cytoplasm. It is lighter in colour, clearer, and with fewer granules (↓) than endoplasm.

endoplasm (*n*) the inner layer of cytoplasm (↑); it is more liquid than ectoplasm (↑) and contains many granules (↓).

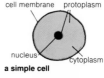

a simple cell

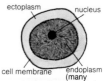

an animal cell

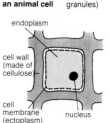

young plant cell

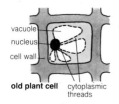

vacuole
nucleus
cell wall

old plant cell   cytoplasmic
threads

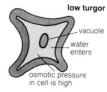

low turgor

vacuole

water
enters

osmotic pressure
in cell is high

vacuole

**full
turgor**

osmotic pressure
= turgor pressure

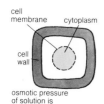

cell
membrane   cytoplasm

cell
wall

osmotic pressure
of solution is
greater than
osmotic pressure
of cell

plasmolysis

nucleus[2] (*n*) a small body, of dense material, covered with a membrane, in the cytoplasm of a cell. It controls all the activities of the cell, and without it, the cell dies. The nucleus contains nuclear sap (a liquid) and chromosomes (p.142). **nuclear** (*adj*).

granule (*n*) a small piece of material, usually hard, separate from other granules, and from other material. **granular** (*adj*).

vacuole (*n*) a space in cytoplasm (↑) enclosed by a membrane and filled with liquid. Many plant cells have a single vacuole which takes up most of the volume of the cell; it is filled with cell sap (a liquid) which has the same osmotic pressure (↓) as the cytoplasm around it. Also a minute space in any tissue. **vacuolar** (*adj*).

osmotic pressure the pressure arising from a solute (p.89) in a solution, which causes water to pass through a membrane (↑) to dilute the solution, provided the solute itself cannot pass through the membrane. The osmotic pressure is the pressure which has to be applied to the solution to stop the movement of water through the membrane. Water passes through a membrane from a solution with a lower osmotic pressure to one with a higher osmotic pressure, until both pressures are equal. The greater the concentration of a solution, the greater the osmotic pressure. (If a **solute** passes through a membrane then diffusion takes place.)

turgor (*n*) the state of a plant cell when fully expanded because of water absorbed (p.163) by its cytoplasm and vacuole; this produces a turgor pressure which keeps the cell wall (↑) rigid. When turgor pressure is equal to the osmotic pressure (↑) of the cytoplasm, no more water enters the cell through the cell membrane (↑). **turgid** (*adj*).

plasmolysis (*n*) the shrinking of the cytoplasm (↑) of a plant cell away from the cell wall (↑) when the cell is put in a solution with a higher osmotic pressure (↑) than that of the cytoplasm. Water passes through the cell membrane (↑) from the vacuole (↑) to the solution and turgor (↑) is lost. **plasmolyse** (*v*).

function (*n*) the work done by any part of a plant
or animal, and also the purpose of that work,
e.g. (a) a function of the nose is to take in air;
(b) a function of the blood is to provide the
body tissues (↓) with oxygen. **function** (*v*).

inhibit (*v*) in animals, to stop or slow down any
function (↑) or action as a result of control by
nerves. **inhibition** (*n*).

physiological (*adj*) concerning the function (↑) of
any part of an organism (p.147) as opposed to its
structure, e.g. a nerve poison has a physiological
effect on the nerves of an animal. **physiology** (*n*).

biological (*adj*) to do with biology.

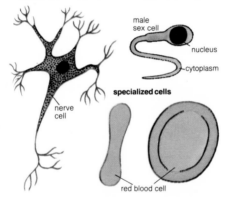

male
sex cell

nucleus

cytoplasm

specialized cells

nerve
cell

red blood cell

specialization (*n*) the change in an organism
(p.147), part of an organism, or a cell (p.138), in
order to perform a particular function or to
improve a function, e.g. (a) the specialization of
cells as seen in nerve cells and muscle cells;
(b) specialization in animals as seen in the
growth of feathers by birds to help in flying.
specialized (*adj*).

tissue (*n*) a mass of cells and the intercellular
material surrounding them all of which perform
the same function, e.g. muscle tissue has the
ability to contract (p.38). The tissue consists
mainly of cells of the same function, e.g. muscle
cells in muscle tissue, but other cells are needed
to bind the tissue together.

organelle (*n*) a specialized part or unit of a cell
with a particular function, (↑), e.g. mitochondria
(↓), ribosomes (↓) are organelles.

mitochondrion (*n*) (*mitochondria*) a very small
granular or rod-shaped body in the cytoplasm
(p.138) of all cells. It uses oxygen and
substances in the cytoplasm to set free energy
for many of the cell's functions.

ribosome (*n*) a very small granular body in the
cytoplasm (p.138) of all cells. Ribosomes
synthesize (p.136) proteins (p.172) from amino
acids (p.172).

plastid (*n*) a very small body, of various shapes,
in the cytoplasm (p.138) of plant cells; some
cells have few, some cells have many. Their
function is to provide colour in plants; some
produce starch (p.174), some protein (p.172)
and some store starch and protein.

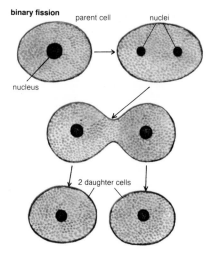

binary fission the division of one cell into two
daughter cells of equal size. The nucleus (p.139)
divides into two, the two nuclei separate, and the
cytoplasm then divides into two and the two
cells are formed.

centriole (*n*) a very small granule just outside the nuclear membrane (p.138). When binary fission (p.141) is about to take place, the centriole divides in two and the process starts of the nucleus dividing in two.

chromosome (*n*) a thread-like body in the nucleus (p.139) of a cell. Each animal and each plant has a particular number of chromosomes, e.g. human beings have 46 chromosomes. All chromosomes are in pairs, so a human being has 23 pairs in the nucleus of every cell in his body, except for sex cells which have half that number through meiosis (↓). The chromosomes in the nuclei determine the particular characteristics of an organism, e.g. height, appearance, colour of hair and eyes, etc. **Sex chromosomes** a pair of chromosomes in the nucleus of a cell which determine the sex of an animal. One sex has a pair of identical (p.94) chromosomes, called X-chromosomes. The other sex has one X-chromosome and either a different chromosone, called a Y-chromosome, or no other chromosome. In mammals (p.150), a Y-chromosome produces a male, i.e. the sex chromosomes are XY; two X-chromosomes produce a female; the sex chromosomes are XX.

meiosis (*n*) the process by which a nucleus (p.139) divides to form sex-cells. The chromosomes of a chromosome pair separate, one going to each side of the nucleus under the control of the centriole (↑). This forms two nuclei, each with half the number of chromosomes, hence the name **reduction division**. The chromosomes then each form two chromosomes and a further division, which is exactly the same as in mitosis (↓), takes place. This results in four nuclei each containing half the number of chromosomes that the nucleus had at the start of meiosis. The cytoplasm divides to form four sex cells with the four new nuclei. **meiotic** (*adj*).

mitosis (*n*) the process by which a nucleus (p.139) divides in two during binary fission (p.141). Each chromosome first forms two chromosomes; these two chromosomes separate, one going to each side of the nucleus under the control of the

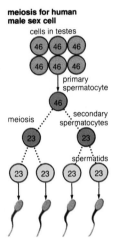

meiosis for human male sex cell

cells in testes

primary spermatocyte

meiosis — secondary spermatocytes

spermatids

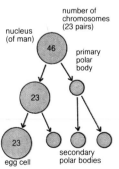

number of chromosomes (23 pairs)

nucleus (of man)

46

primary polar body

23

secondary polar bodies

egg cell

meiosis for human female sex cell

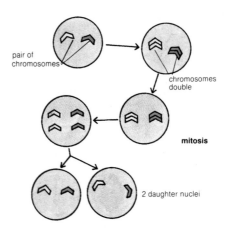

pair of
chromosomes

chromosomes
double

mitosis

2 daughter nuclei

centriole (↑), while the nuclear membrane (p.138)
disappears. A new nuclear membrane is formed
round each group of chromosomes and two new
nuclei are formed. The process usually takes
between 30 and 180 minutes. **mitotic** (*adj*).

polar body a very small cell containing a nucleus,
but almost no cytoplasm (p.138) at all, formed
during the formation of a female sex cell. In the
first division of meiosis, one egg-cell (a female
cell) and a primary polar body are formed.
Each of these two cells then divides mitotically
(↑) forming one egg-cell and three polar bodies,
see diagram. The polar bodies die leaving the
egg-cell.

Protozoa (*n*) a group of animals which are
different from all other animals because they
consist of only one cell, and thus are
microscopically (p.61) small. Some protozoa
show a relation with simple plants. Protozoa
live under almost all conditions found in nature,
and are important in ecological (p.227) relations.

Amoeba (*n*) a group of protozoa (↑) of irregular
shape. An amoeba continually changes its shape
by movement of its cytoplasm. It feeds by
closing pseudopodia (p.144) round a food particle
and absorbing the food inside a vacuole (p.139).
amoebic (*adj*), **amoeboid** (*adj*).

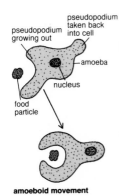

pseudopodium
growing out

pseudopodium
taken back
into cell

amoeba

nucleus

food
particle

amoeboid movement

pseudopodium (*n*) (*pseudopodia*) a part of a cell (in some protozoa) formed by the cytoplasm (p.138) pushing out temporarily an arm of irregular shape. Pseudopodia are used in locomotion (p.194) and in feeding; they are formed, are used for these purposes, and then are taken back into the cell. This kind of locomotion is called **amoeboid movement**.

structure (*n*) (1) a part of an organism which is complete in itself, but which has no shape that can be described, i.e. it is not a surface, a cavity or a vessel. Examples of structures are: a tooth, an ear, a tail. (2) the arrangement of cells and tissues (p.140) in a part of an organism, e.g. the structure of a muscle. **structural** (*adj*).

cilium (*n*) (*cilia*) a thread-like structure (↑) which together with many others grows from the surface of a cell. Cilia beat, one after the other, in a regular way; they either push water past a cell, or push the cell through water, *see diagram*. **ciliary** (*adj*).

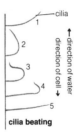

cilia beating

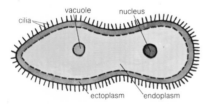

paramecium

Paramecium (*n*) a group of protozoa (p.143) of a shoe-like shape. The surface of a Paramecium is covered with cilia (↑) and ciliary beating drives the animal through the water.

flagellum (*n*) (*flagella*) a long, thread-like structure, longer than a cilium (↑), with a wave-like movement for beating in water; it is used for locomotion (p.194). One or two may be found on some unicellular organisms but bacteria (↓) can have groups of flagella called **tufts** (*n*). Organisms with flagella are described as **flagellate** (*adj*).

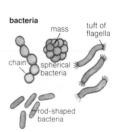

bacteria

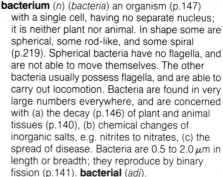

influenza virus

protein coat

genetic material for reproduction

cytoplasm of bacterium

cell wall

bacteriophage attacking a bacterium

bacteriophage

bacterium (*n*) (*bacteria*) an organism (p.147) with a single cell, having no separate nucleus; it is neither plant nor animal. In shape some are spherical, some rod-like, and some spiral (p.219). Spherical bacteria have no flagella, and are not able to move themselves. The other bacteria usually possess flagella, and are able to carry out locomotion. Bacteria are found in very large numbers everywhere, and are concerned with (a) the decay (p.146) of plant and animal tissues (p.140), (b) chemical changes of inorganic salts, e.g. nitrites to nitrates, (c) the spread of disease. Bacteria are 0.5 to 2.0 μm in length or breadth; they reproduce by binary fission (p.141). **bacterial** (*adj*).

virus (*n*) an agent of disease which is not usually considered to be living, as some viruses can be crystallized (p.110), but as all viruses can reproduce inside a living cell, they can be considered to be alive. Bacteria can be grown on suitable substrates (p.146) outside the body, but viruses cannot. The diameter of a virus is between 0.5 μm and 0.03 μm, i.e. much smaller than that of bacteria. **viral** (*adj*).

bacteriophage (*n*) a virus that attacks and destroys bacteria.

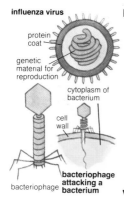

sporangium

spores

hypha

fungus

fungus (*n*) (*fungi*) a plant containing no chlorophyll (p.159). It consists either of a single cell, or of cellular tube-like threads; it feeds on dead or living organisms (p.147) and is an agent of decay (p.146) or disease. Fungi reproduce by spores (p.146). **fungal** (*adj*).

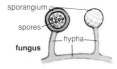

sporangia

mycelium

hypha entering plant tissues

hypha (*n*) a thread-like part of a fungus (↑), e.g. as seen growing on old bread. It grows at the end, or by branches; it is tube-like, and filled with cytoplasm (p.138), which is not separated by cell walls or membranes; nuclei are present in the cytoplasm. **hyphal** (*adj*).

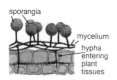

sporangium

mushroom (a fungus)

spores

ground level

mycelium

mycelium (*n*) (*mycelia*) the mass of hyphae (↑) which forms all parts of the fungus, other than the reproductive organs, *see diagram*.

sporangium (*n*) (*sporangia*) a structure, covered by a wall, containing spores (p.146). It is formed at the end of a hypha (↑) growing up from a mycelium (↑).

spore (*n*) a reproductive cell, or a group of cells, with a wall around it, formed by certain plants, particularly ferns and fungi; also formed by some bacteria and some protozoa (p.143). Plant spores are produced by vegetative reproduction (p.213) in very large numbers, and from each spore, a new plant can grow under favourable conditions; the spores differ from seeds in not having an embryo (p.154). Spores of bacteria and protozoa are formed from a single organism (↓) when conditions are unfavourable.

cyst (*n*) a bag-like cover formed round a resting protozoan, e.g. an amoeba forms a cyst when conditions are unfavourable. **cystic** (*adj*).

yeast (*n*) a fungus consisting of single cells; it multiples by budding, i.e. small cells grow out from the mother cell, *see diagram*. Yeasts produce enzymes (p.167) which cause the decomposition of starch and sugars to alcohol (ethanol) and carbon dioxide. There are many different kinds of yeasts, each acting on a different substrate (↓).

mould (*n*) a grey or white growth of a fungus (p.145) on the surface of a living, or specially a dead, organism (↓); it causes decay of the organism, e.g. mould growing on old bread. **mouldy** (*adj*).

substrate (*n*) (1) the ground, or other object, on which animals live; (2) the substance on which microscopic organisms (↓) live and feed; (3) the particular substance on which an enzyme (p.167) acts.

decay (*n*) the chemical processes, caused by bacteria (p.145) and fungi (p.145), which take place in animal and plant substances after death. Decay breaks down the tissues and chemical compounds until only animal bones are left, while plants completely disappear. **decay** (*v*).

Penicillium (*n*) a mould (↑) from which penicillin (↓) is produced.

penicillin (*n*) a substance which prevents the growth of bacteria; it is an antibiotic (p.240). Fungi (p.145) produce substances which prevent other organisms growing near them, and penicillin is such a substance.

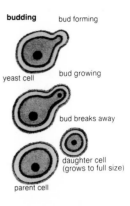

budding   bud forming

yeast cell   bud growing

bud breaks away

daughter cell
(grows to full size)

parent cell

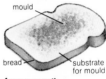

mould

bread   substrate
for mould

fungus growth

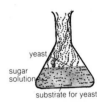

yeast

sugar
solution

substrate for yeast

fermentation of substrate

vertebrates

mammal
hair
mammary
glands
young
born alive
warm-
blooded

bird
feathers
beak
wings
warm-
blooded

reptile
scales cold-blooded

amphibian
smooth skin
four legs
cold-blooded

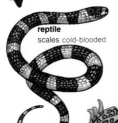

organism (*n*) a plant, an animal, anything that lives, such as a tree, a man, a bacterium.

nature (*n*) (1) all objects, organisms and forms of energy considered as a system (p.162) but excepting objects made by man, e.g. mountains, rivers, the atmosphere, plants, animals, are all parts of nature. (2) all the properties (p.27) and characteristics (↓) which determine the whole behaviour of an object, organism, or form of energy, e.g. (a) the nature of a metal depends on its properties; (b) the nature of a fish depends on its characteristics.

characteristic (*n*) any part, shape, way of behaving, by which an organism, or group of organisms, can be recognized, e.g. (a) hair on the body is a characteristic of mammals; (b) feathers are a characteristic of birds; (c) green-coloured leaves are a characteristic of most plants; (d) walking on two legs is a characteristic of man. Objects can have their characteristics, e.g. the conduction of heat is a characteristic of metals. **characteristic** (*adj*).

classify (*n*) to put objects or organisms into named groups, chosen by examining the properties (p.27) or characteristics (↑) of the object or organism and of the group, e.g. (a) an animal has a beak, feathers, two legs and two wings, and these are the characteristics of a bird, so the organism is classified as a bird; (b) a substance is hard, is a good conductor of heat and electric current, reacts with acids to form a salt, and these are the characteristics of a metal, so the substance is classified as a metal. **classification** (*n*).

fish
scales
fins
cold-blooded

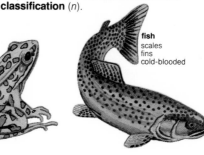

species (*n*) (*species*) the smallest group in the classification (p.147) of organisms (p.147). Members of a species can be parents of young, who can in turn be parents. Pairs from different species usually cannot produce young; if they do, the young cannot themselves reproduce (p.209), e.g. two horses are members of the same species and can reproduce young horses; a horse and a donkey can produce a mule, but mules cannot reproduce, as horses and donkeys belong to different species.

animal (*n*) a living being which cannot make its food from simple inorganic (p.116) substances, but obtains its food from plants or other animals. Most animals use locomotion (p.194) to help them find food.

vertebrate (*n*) an animal possessing a line of bones, a backbone, which supports and protects its body; each bone is called a vertebra (p.193). The main groups of vertebrates are: fishes, amphibians (↓), reptiles (p.150), birds and mammals (p.150).

invertebrate (*n*) an animal possessing no backbone; some are completely soft, e.g. worms, and some have a hard outer covering, e.g. insects, crabs.

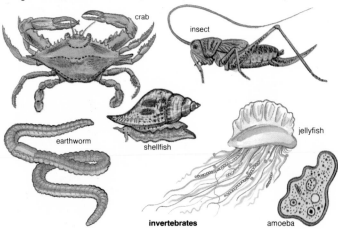

crab

insect

earthworm

shellfish

jellyfish

invertebrates

amoeba

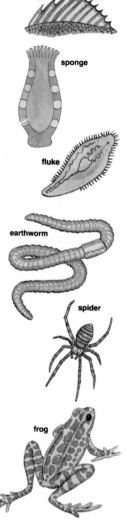

fin

sponge

fluke

earthworm

spider

frog

appendage (*n*) a relatively large structure attached to the main body of an animal, e.g. a leg, an arm, a tail, are all appendages. **Paired appendages**, with one on each side of the body, are common in many animals, e.g. a pair of arms. **appendicular** (*adj*).

fin (*n*) a thin, flat structure consisting usually of many small bones joined by a continuous piece of skin; found on all fish and some other animals which live in water. Fins are used to control the direction of motion, and also are used in locomotion (p.194).

sponge (*n*) an animal, living in water, consisting of many cells, but having no nerves, and not able to move. It has tissues, but no organs. Water is drawn in through small holes by the movement of flagella (p.144) and passed out through larger holes; food particles are caught from the water.

fluke (*n*) a kind of very small worm with a flat body; some flukes can be as long as 1 cm. Fully grown flukes live in the liver, gut, lung or blood vessels of vertebrates (↑), with different kinds of fluke in each place. Flukes can cause serious diseases, especially in man.

worm (*n*) a general name for many different kinds of invertebrates (↑) having a long, thin, soft body, and no appendages (↑). It is not a biological name for an animal.

vermiform (*adj*) describes a structure with a shape like a worm.

arthropod (*n*) one of a group of animals which possess (a) a hard outer covering to their body instead of bones to support the body, (b) paired legs consisting of jointed tube-like structures. Some live in water, e.g. crabs; others live on land, e.g. centipedes, spiders, insects (p.151).

amphibian (*n*) one of a group of animals which are vertebrates (↑) and possess (a) four legs, each with five digits (fingers or toes), (b) a moist, smooth skin with no scales; their eggs are not protected by a shell. Amphibians live partly in water and partly on land. Examples of amphibians are frogs and toads. **amphibious** (*adj*).

reptile (*n*) one of a group of animals which are vertebrates and possess (a) horny skins or skin covered with horny plates, (b) four legs, or no legs; their eggs (p. 217) are protected by a hard, horny shell. Some reptiles live in water, e.g. turtles and crocodiles, but breathe air, other reptiles live on land, e.g. lizards, tortoises and snakes. **reptilian** (*adj*).

bird (*n*) one of a group of animals which are warm-blooded (p.188) vertebrates and possess (a) skin covered in feathers, (b) legs covered in scales, (c) beaks without teeth; their young are hatched from eggs (p.217) with a large yolk and a hard shell, e.g. doves, sparrows, eagles.

mammal (*n*) one of a group of animals which are warm-blooded (p.188) vertebrates and possess (a) skin covered in hair, (b) on females, mammary glands producing milk; their young are born alive. Some mammals live in the sea, e.g. whales and seals, others live on land, e.g. horses, lions, monkeys, men. **mammalian** (*adj*).

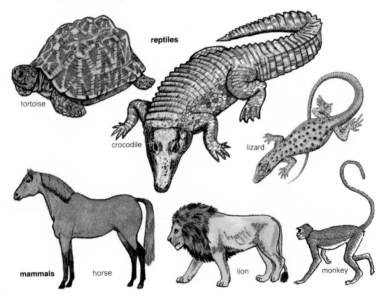

reptiles

tortoise

crocodile

lizard

mammals   horse   lion   monkey

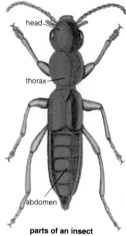

parts of an insect

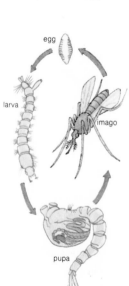

life cycle of a mosquito

insect (*n*) a fully grown arthropod (p.149) with its body divided into three separate parts, head, thorax and abdomen, *see diagram*. It possesses (a) three pairs of jointed legs, (b) one pair of antennae, (c) two pairs of wings (in many, but not all, insect groups). It has a life cycle (↓) in which it changes its appearance and way of life, e.g. flies, bees, moths.

larva (*n*) (*larvae*) one part of the life cycle (↓) of an insect, and of other invertebrates (p. 148); it comes after an egg (p. 217). An egg hatches (p.217) and a larva comes out. A larva is quite different in appearance from the fully grown insect, usually being worm-like, *see diagram*; it cannot reproduce. It changes into a pupa or a fully grown insect or other invertebrate. **larval** (*adj*).

pupa (*n*) (*pupae*) a resting stage (p.152) in the life of some insects; it no longer moves or feeds. It is enclosed in a case, and great changes take place in the structure of its body, e.g. a caterpillar changes into a butterfly. **pupate** (*v*), **pupal** (*adj*).

cocoon (*n*) a protective cover made by a larva (↑) for the pupa (↑); also a protective cover for eggs made by some invertebrates, e.g. earthworms make a cocoon for their eggs. The cocoon of the silkworm is the source of silk.

imago (*n*) a fully grown insect into which a pupa changes. It can reproduce and the female forms eggs, which starts the life cycle (↓) of the insect.

metamorphosis (*n*) the change of form from a larva (↑)to an imago (↑). *Complete metamorphosis* is the change of an insect from larva to pupa to imago; *incomplete metamorphosis* is the change of an insect from larva to imago, with the larva being similar in appearance to the imago, e.g. a cockroach undergoes incomplete metamorphosis.

life cycle the changes, one after another, through which a plant or animal passes, e.g. the changes which take place from the production of an egg to the death of the organism. For many insects this is egg-larva-pupa-imago-death. It is represented in the diagram with the imago producing an egg for the next generation.

moult (*v*) to lose an outer covering such as hair, feathers or skin at regular times, e.g. a cat moults and loses hair in hot weather. In larvae (p.151) which undergo incomplete metamorphosis (p.151), the hard outer cover splits and the larva draws itself out, with a new, soft cover which hardens in the air.

life history the changes which take place in an organism, from the production of an egg to death. Also, the changes in one stage (↓) of a life cycle (p.151) of an organism, e.g. the life history of a larva. The life cycle of a mammal is the same as its life history, but the life cycle of an insect consists of the life histories of each stage.

stage (*n*) one part, out of two or more parts, of a course of work, or a particular process or the life of an organism (p.147), e.g. (a) the stages of school work; (b) the stages in a process of distillation, i.e. boiling the mixture – collecting the distillate; (c) the stages in the life of a mosquito: egg-larva-pupa-imago.

maggot (*n*) a worm-like larva of an insect, e.g. the maggot of a house-fly.

caterpillar (*n*) the larva of a butterfly (↓) or moth. Its body is soft, long and round, and is divided into fourteen parts, called segments. The first segment is the head, the next three are thoracic (p.189) segments and the last ten are abdominal (p.162) segments. There are a pair of legs on each thoracic segment and a pair of structures on each abdominal segment called prolegs. The body is usually covered in fine hairs. A caterpillar eats leaves and plants, and some caterpillars are serious pests (p.231).

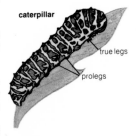

caterpillar
true legs
prolegs

chrysalis (*n*) the pupa of a butterfly (↓), a moth and of some other insects.

butterfly (*n*) an insect (p.151) with a short, fat body divided into fourteen segments, the same as in a caterpillar (↑). Two pairs of large wings are fixed to the 2nd and 3rd thoracic segments, and a pair of jointed legs to each thoracic segment. The wings and body are covered with small scales. Butterflies fly by day and feed on nectar (p.213) from flowers. Moths are similar to butterflies, but fly by night.

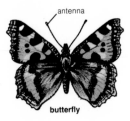

antenna
butterfly

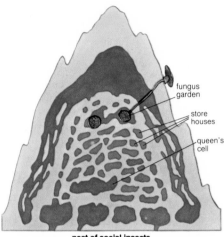

nest of social insects

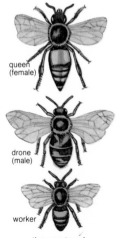

queen
(female)

drone
(male)

worker

**three castes of
honey bee**

social insect an insect which lives in a group
with a division of work between the different
castes (↓), e.g. bees and termites are social
insects.

society (*n*) a group of social insects.

solitary insect (*n*) an insect that lives alone and
does not belong to a society.

caste (*n*) a particular kind of social insect (↑)
with a structure and a function for a special kind
of work, e.g. honey bees have three castes for
the division of labour in a hive (↓): queens (↓),
workers (↓) and drones (↓).

queen (*n*) a female social insect (↑) that produces
eggs. There is only one queen in a hive (↓) of
honey bees; she is larger than the members of
the other two castes and has a pointed abdomen.

drone (*n*) a male bee which does no work; it
mates with a queen.

worker (*n*) a female social insect not able to
produce eggs; it is smaller than either a queen
or a drone and does all the work in a hive (↓).

hive (*n*) the place in which bees live, a kind of
very large nest; also the bees gathered together
in that place.

seed (*n*) the small body, produced by flowering plants, from which a new plant grows. It is a product of sexual reproduction (p.209) from the union of male and female parts of flowers (p.211) and consists of several different structures.

testa (*n*) (*testae*) the hard, outer covering of a seed. It does not let water pass through; when dry it does not let oxygen pass, but when it is wet, oxygen passes through.

micropyle (*n*) a very small hole in the testa (↑) of a seed. Water enters a seed through its micropyle, then the seed begins to germinate (p.156).

hilum (*n*) a scar, i.e. a mark, on the testa (↑) of a seed; it shows where the seed was fixed to the parent plant.

embryo[1] (*n*) a young plant in a seed; it is formed from an ovule (p.211) in a flower. An embryo is the part of a seed which grows into a new plant. **embryonic** (*adj*).

plumule (*n*) the part of an embryo which grows to form the main stem of a new plant.

radicle (*n*) the part of an embryo which grows to form the main root of a new plant.

shoot (*n*) the stem of a young plant which grows from the plumule (↑); leaves grow from the shoot.

endosperm (*n*) food material for the use of an embryo (↑) in a seed. Not all seeds have an endosperm, but where they do, the endosperm surrounds the embryo. **endospermous** (*adj*).

cotyledon (*n*) a simple leaf, usually lacking chlorophyll (p.159), forming part of the embryo of a seed. Flowering plants have one or two cotyledons. In some plants, e.g. legumes (↓), the cotyledons have a food store for the growing embryo; in other plants, e.g. maize, the cotyledon takes in food from the endosperm and passes it to the embryo. In many plants, the cotyledons appear above ground, produce chlorophyll, and make food for the plant. **cotyledonous** (*adj*).

dormant (*adj*) describes any organism that is resting and not growing. Seeds are dormant in soil until the temperature is high enough.

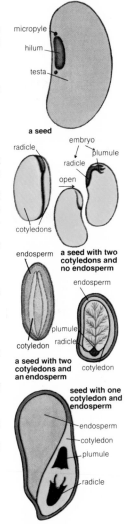

a seed

a seed with two cotyledons and no endosperm

a seed with two cotyledons and an endosperm

seed with one cotyledon and endosperm

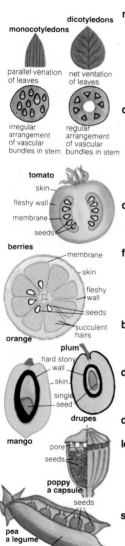

monocotyledons

dicotyledons

parallel venation of leaves

net ventation of leaves

irregular arrangement of vascular bundles in stem

regular arrangement of vascular bundles in stem

tomato

skin

fleshy wall

membrane

seeds

berries

membrane

skin

fleshy wall

seeds

succulent hairs

orange

plum

hard stony wall

skin

single seed

drupes

mango

pore

seeds

poppy a capsule

seeds

pea a legume

pod

monocotyledon (*n*) a flowering plant with one cotyledon in its seed. All monocotyledons have (a) parallel venation (p.158); (b) flower parts (p.211) in groups of threes; (c) vascular bundles (p.160) irregularly arranged. Most monocotyledons are small plants, but a few are large, e.g. bananas, palm trees. **monocotyledonous** (*adj*).

dicotyledon (*n*) a flowering plant with two cotyledons in its seed. All dicotyledons have (a) net venation (p.158); (b) flower parts (p.211) in groups of four or five; (c) vascular bundles (p.160) in the form of a ring round the centre of a stem. All trees except palms are dicotyledons; there are many more species of dicotyledons than of monocotyledons. **dicotyledonous** (*adj*).

dispersal (*n*) the process by which seeds get to places away from the parent plant. The means of dispersal are wind, water and animals. **disperse** (*v*).

fruit (*n*) a body, formed from the ovary (p.211) of a plant, containing and protecting seeds. Common kinds of fruits include berries, drupes, capsules, legumes (↓). When describing biological structures, fruit has a different meaning from that in everyday use.

berry (*n*) a succulent (↓) fruit containing many seeds, with an outer skin, a thick fleshy wall, and an inner membrane round the inside part which contains seeds, e.g. tomato, orange, guava.

drupe (*n*) a succulent (↓) fruit containing a single seed with an outer skin, a thick fleshy wall, and a hard woody layer round the seed, e.g. plum, mango, cherry.

capsule (*n*) a dry fruit which splits open to set free seeds, e.g. poppy.

legume (*n*) (1) a dry fruit, also known as a pod, with a long narrow skin consisting of two halves with seeds inside. The skin splits along the join of the two halves and the seeds fall out, e.g. pea, bean. (2) any plant belonging to the pea or bean family.

succulent (*adj*) soft, thick and containing a lot of water, e.g. cacti and many fruits that are eaten are succulent.

prolific (*adj*) producing seeds, spores, eggs, young, in large numbers.

germinate (*v*) to start to grow, of seeds and spores. The growth produces a new plant, called a seedling (↓). In order for a seed, or spore, to germinate, the conditions round it must be suitable. **germination** (*n*).

seedling (*n*) a newly formed organism growing from a seed. The cotyledons (p.154) help to provide food during this stage of growth; they die when the seedling becomes a plant.

epigeal (*adj*) having growth above ground level; epigeal seedlings have their cotyledons (p.154) above ground during growth. Most dicotyledons (p.155) are epigeal.

hypogeal (*adj*) having growth below ground level; hypogeal seedlings have their cotyledons (p.154) below ground level during growth. Most monocotyledons (p.155) are hypogeal.

epicotyl (*n*) the part of a seedling above the cotyledons (p.154) and below the first foliage (p.158) leaves. It grows bent in a half circle to protect the plumule (p.154) during germination; when the plumule is above ground, it straightens and becomes the first part of the stem.

hypocotyl (*n*) the part of a seedling below the cotyledons (p.154) and above the root. It is the stem of the embryo seed plant.

permeable (*adj*) allows molecules (p.103) and ions (p.101) to pass through, e.g. a permeable membrane allows molecules or ions, but does not allow fluids, to pass through; such a membrane is *impervious* to fluids but permeable to solvents, solutes and ions.

impermeable (*adj*) does not allow molecules or ions to pass, i.e. it is not permeable (↑).

impervious (*adj*) does not allow fluids to pass, but may be permeable (↑).

plant (*n*) a living object that makes its own food from simple inorganic substances which are gases or in solution; it usually has no power of locomotion (p.194).

stem (*n*) the part of a plant that bears leaves and buds (p.213). Most stems grow above ground, but some stems grow below ground level.

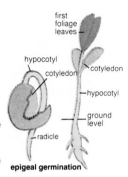

first foliage leaves

hypocotyl

cotyledon

cotyledon

hypocotyl

ground level

radicle

epigeal germination

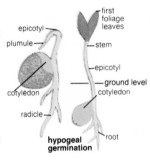

first foliage leaves

epicotyl

plumule

stem

epicotyl

ground level

cotyledon

cotyledon

radicle

root

hypogeal germination

branch

stalk

stem

internode

node

prop roots (maize)

adventitious roots

clasping roots (fig)

roots

swollen tap root (carrot)

aerial roots (orchid)

roots storing food

swollen fibrous roots (tapioca) (cassava)

stalk (*n*) a structure without branches, supporting one or several objects at the top. It may be found in objects, or in plants, e.g. a straight, unbranched stem with a flower or flowers, or in animals, e.g. the eye-stalk of a crab.

node[2] (*n*) the part of a stem (↑) from which leaves (p.158) grow. Adventitious roots (↓) may also grow from nodes. **nodal** (*adj*).

internode (*n*) the part of a stem (↑) between two nodes (↑); no leaves grow on it.

lenticel (*n*) a small, raised hole, usually elliptical, formed in woody (p.161) stems; it allows gases to enter and leave the stem.

root (*n*) the part of a plant that grows downwards into the earth; it differs from a stem in not having leaves or buds. Its functions are: (1) to hold the plant firmly in the ground; (2) to absorb water from the earth; (3) to absorb inorganic salts (p.115) from the earth. A root-cap protects the end of the root.

tap root the main root, growing straight downwards, of dicotyledon (p.155) plant; other roots branch from it. It is formed from the radicle (p.154) of the seed. The tap roots of some plants become swollen with stored food, e.g. carrot.

adventitious root a root growing from any node (↑) on a stem. There are several kinds of adventitious root, e.g. a prop root (↓) is an adventitious root. Adventitious roots also grow from bulbs.

prop root a root growing from a node (↑), on a stem, down into the earth; it supports the stem.

fibrous root one of many branching roots growing from the bottom of a stem, usually with no tap root in the system of roots. Monocotyledons (p.155) have fibrous roots, and no tap root.

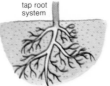

tap root system

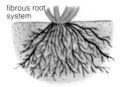

fibrous root system

root hair a small, hair-like growth from a root, with a very thin cell wall. Large numbers of root hairs grow near the end of the root; their function is to take in water and inorganic salts from the earth. They have a short life, as they die when the root grows downwards.

piliferous layer the part of the surface of a root bearing root hairs; it covers only part of the root, near the end. A layer of cells near the outside of the root forms the piliferous layer.

leaf (*n*) (*leaves*) a flat structure, usually green, growing on a stalk from the node (p.157) of a stem or branch of a plant. The important functions of a leaf are photosynthesis (↓) and transpiration (↓). **leaflet** (*n*) a small leaf.

margin (*n*) (1) the edge of a leaf, which can have various shapes depending on the species of plant, *see diagram.* (2) the edge, and the flat space near the edge, of an object, e.g. the margin of an insect's wing. **marginal** (*adj*).

petiole (*n*) the stalk of a leaf; it grows from a node (p.157).

venation (*n*) the arrangement of small tubes seen in a leaf; the tubes are called *veins*, and the veins act as a support and conduct water and nutrients (p.171) for the leaf. There are two kinds of venation, parallel and net (see p.158).

foliage (*n*) the mass of leaves on a plant. A foliage leaf is the common kind of leaf, different from special leaves such as cotyledons (p.154) and other leaf-like parts of flowers.

axil (*n*) the angle between the upper side of a petiole and the stem of the plant. **axillary** (*adj*).

modified (*adj*) changed in shape and function. A modified leaf has a different shape from a common leaf, and has a different function, e.g. (1) leaf tendril, in which the leaf is modified to form a long, thin, thread-like shape (tendril) which twists round an object in order to support the plant, (2) leaf spine, such as a spine on a cactus which is a small, sharp-pointed, modified leaf.

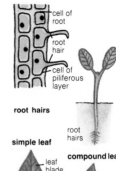

root hairs

simple leaf

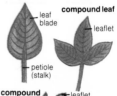

compound leaf

leaf blade

leaflet

petiole (stalk)

compound leaf

leaflet

leaf blade

different kinds of leaves

margin

margins of leaves

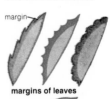

axil

leaf

petiole

stem   leaf

vein

venation

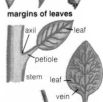

leaf tendrils (gloriosa)

tendril for clasping

deciduous (*adj*) losing all leaves at a certain season of the year, e.g. many deciduous trees lose their leaves in autumn.

evergreen (*adj*) bearing leaves at all times, e.g. pine trees are evergreen as they never lose all their leaves at one time.

chlorophyll (*n*) a green substance which gives leaves their colour. Chlorophyll takes in energy from sunlight, and a plant uses this energy to make food for itself (the process known as photosynthesis (↓)).

photosynthesis (*n*) in green plants, sunlight is absorbed by chlorophyll (↑) contained in plastids (p.141). This energy is used to synthesize (p.136) organic (p.131) compounds from carbon dioxide and water. These carbohydrates (p.173) are mainly stored as starch. A small part is used as the substrate (p.146) for respiration. **photosynthetic** (*adj*)

guard cells and stomata

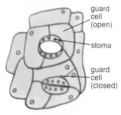

guard cell (open)

stoma

guard cell (closed)

stoma (*n*) (*stomata*) a very small hole in the surface of a leaf; it has two guard cells (↓) round it. Oxygen and carbon dioxide from the air enter through the stomata; oxygen, carbon dioxide and water vapour leave through the stomata. See transpiration (↓) and respiration (p.191).

guard cell one of a pair of bean-shaped cells round a stoma (↑). The cell-wall facing the stoma is thick. When the turgor (p.139) pressure is high, the stoma is open as the guard cell wall is curved, but when the turgor pressure is low, the stoma is closed.

root pressure the pressure which forces water up from a root into the stem of a plant. The osmotic pressure of sap (p.160) in the cells of the root causes root pressure.

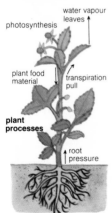

photosynthesis

water vapour leaves ↑

plant food material ↓

transpiration pull ↗

plant processes

root pressure ↑

transpiration (*n*) the loss of water vapour by plants, mainly through their stomata (↑). This action helps root pressure by drawing water up the stem, and a plant thus increases the amount of water passing up its stem. **transpire** (*v*), **transpiratory** (*adj*).

potometer (*n*) a piece of apparatus (p.88) for measuring the rate of transpiration (↑), or the rate of movement of the transpiration stream in the stem of a plant.

xylem (*n*) the part of the vascular system (p.178)
  containing pipe-like conducting vessels and
  wood (p.161). The vessels conduct water from
  the root to all parts of the plant. The wood
  contains dead cells and provides support for the
  plant stem. Water passes up the xylem vessels
  pushed by root pressure (p.159) and pulled by
  transpiration (p.159).

phloem (*n*) the part of the vascular system (p.178)
  containing tube-like conducting vessels in which
  dissolved food material passes from the leaves
  to all parts of a plant.

stem of young dicotyledon plant

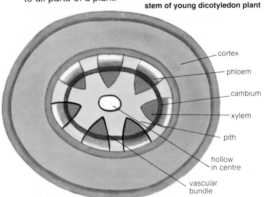

- cortex
- phloem
- cambium
- xylem
- pith
- hollow in centre
- vascular bundle

vascular bundle a bundle of xylem (↑) and
  phloem (↑) vessels passing from the end of a
  root to the end of a stem. The arrangement of
  vascular bundles in monocotyledons (p.155) is
  different from that in dicotyledons (p.155).
  Vascular bundles pass into leaves, where they
  form the veins, see venation (p.158).

sap (*n*) the liquid in a plant; it can be seen when a
  stem or a root is cut.

cortex (*n*) (1) the outer layer of a plant, surrounding
  the vascular system (p.178), *see diagram*. The
  cortex is covered by the cells of the outside
  surface of a plant. (2) the outer part of any structure, e.g. the cortex of a kidney, **cortical** (*adj*).

pith (*n*) the central sponge-like part of the stem of a
  plant; it helps in the storage of plant food.

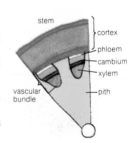

- stem
- cortex
- phloem
- cambium
- xylem
- pith
- vascular bundle

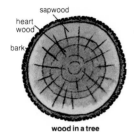

wood in a tree

cambium (*n*) a layer of cells between the xylem and the phloem of dicotyledons (p.155). The cells are actively dividing by binary fission (p.141) forming xylem on one side and phloem on the other; their action makes the stem thicker as the plant grows. Monocotyledons do not possess a cambium.

wood (*n*) the hard part of a stem formed from plant cells whose cellulose walls have been made stronger by a deposit of **lignin.** As the cells get older, they lose all cytoplasm, and only act as a support for the plant. **Heartwood** contains old cells which no longer conduct water or have any cytoplasm. **Sapwood** contains xylem vessels and is not as strong as heartwood. **woody** (*adj*).

tropism (*n*) the tendency of a plant to have curved growth under the effect of its outside conditions, e.g. light, water, gravity. **tropic** (*adj*).

phototropism of stems

phototropism (*n*) a tropism (↑) caused by light. Plant stems become curved so that the plant grows towards the light; this is a *positive tropism.* Some roots grow away from the direction of light; they are *negatively phototropic.* Leaves are always positively phototropic. **phototropic** (*adj*).

hydrotropism (*n*) a tropism (↑) caused by the presence of water. Roots tend to grow towards water in a soil; they show positive hydrotropism. **hydrotropic** (*adj*).

geotropism (*n*) a tropism (↑) caused by gravity. The main stems of a plant show negative geotropism as they grow upwards. The main roots of a plant are positively geotropic as they grow downwards. **geotropic** (*adj*).

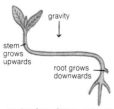

geotropism of stem and root

organ (*n*) a part of a plant or animal which has a definite structure and a particular function (p.140), e.g. (a) the stomach of an animal has a structure and a particular function to help in digesting (p.166) food; (b) the leaf of a plant has a definite structure and a particular function.

system² (*n*) all the organs (↑) and the tissues (p.140) concerned with a particular function of an organism, e.g. the nervous system, which includes nerves, brain and other structures, all concerned with the senses and body activities.

gland (*n*) an organ which takes water and substances from blood, and makes particular chemical substances needed by the organism (p.147). The compounds are passed to the outside of the gland in a solution, called a secretion (↓). **glandular** (*adj*).

secrete (*v*) to pass out from a cell a liquid containing substances made by the cell for use in other parts of the organism. In a gland (↑) special cells make the liquid. Secretion is both the process and the name of the liquid secreted. **secretion** (*n*), **secretory** (*adj*).

duct (*n*) a short pipe through which secretions (↑) leave a gland.

gut (*n*) the part inside an animal which changes food into chemical substances which are used by the body. A simple gut is seen in **hydra**, *see diagram*; it is a pipe with an opening at one end only. The gut in a human being has several organs and is a big system of pipes, glands and organs.

abdomen (*n*) a hollow part of the body of an animal which contains most of the gut (↑), e.g. the abdomen of an insect, or a man. **abdominal** (*adj*).

viscera (*n.pl.*) the organs in the abdomen, together with the heart and the lungs.

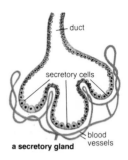
duct
secretory cells
blood vessels
a secretory gland

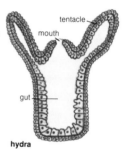
tentacle
mouth
gut
hydra

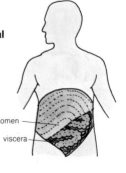
abdomen
viscera

absorb (*v*) to take in a liquid or a gas, e.g. (a) a brick absorbs water; (b) a plant absorbs dissolved mineral salts through its root hairs; (c) a leaf absorbs carbon dioxide through its stomata. The process by which the liquid or gas is absorbed, can be osmosis, diffusion, capillary attraction. **absorption** (*n*), **absorptive** (*adj*).

assimilate (*v*) to take in a material and to change it so that it becomes part of the thing it joins, e.g. when an organism assimilates food, the chemical substances are changed to become part of the body of the organism.

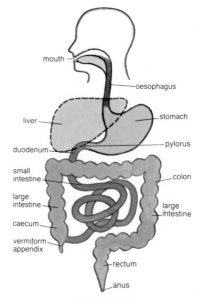

mouth

oesophagus

liver

stomach

duodenum

pylorus

small intestine

colon

large intestine

large intestine

caecum

vermiform appendix

rectum

anus

alimentary canal

alimentary canal in all animals except very simple ones, a long tube, with an opening at each end, concerned with the digestion (p.166) and absorption (p.163) of food. Food enters at the mouth, passes through the alimentary canal, and the unassimilated material leaves the anus (p.165).

oesophagus (*n*) the part of the alimentary canal
between the throat and the stomach. Food is
passed along it by peristalsis (p.169).

stomach (*n*) the part of the alimentary canal after
the oesophagus, where the canal widens to form
a bag-like structure. It has walls made of muscle,
which press and turn the food into a liquid mass.
A secretion (p.162) from the stomach walls acts
on the food; it is called gastric juice (p.166).

pylorus (*n*) the place where the stomach (↑) and
the duodenum (↓) join; it is closed by a strong
circular muscle, which opens and allows food to
pass through after the food has been sufficiently
mixed with gastric juice (p.166). **pyloric** (*adj*).

intestine (*n*) the part of the alimentary canal
between the stomach and the anus (↓). In
vertebrates (p.148) absorption of food takes
place in the intestine. In reptiles, birds and
mammals, the intestine is divided into a small
intestine and a large intestine; the small intestine
is smaller in diameter, but longer than the large
intestine. **intestinal** (*adj*).

small intestine a long tube, with walls of muscle,
between the stomach and the large intestine.
The digestion (p.166) of food is completed and
the absorption of food takes place in this part of
the alimentary canal. The diameter of the small
intestine is about 37 mm.

duodenum (*n*) the part of the small intestine,
joined to the stomach (↑). It is joined to the
liver and to the pancreas by ducts (p.162).
Digestive (p.166) juices are very active in the
duodenum. **duodenal** (*adj*).

ileum (*n*) in mammals, the part of the small intestine before the large intestine.

villus (*n*) (*villi*) a small, finger-like structure on
the inside wall of the small intestine. In human
beings, a villus is about 1 mm long and there are
very large numbers of them; they increase the
absorptive (p.163) surface of the intestine. Each
villus contains blood vessels and the absorbed
food material is carried away by the blood. The
capillaries (p.179) surround a **lacteal**; the lacteal
absorbs fats, which are carried away in lymph
(p.182).

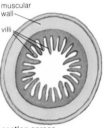

muscular wall
villi

**section across
small intestine**

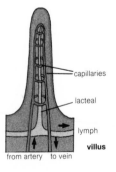

capillaries

lacteal

lymph

villus

from artery   to vein

large intestine a broad tube with muscular walls, connected to the end of the small intestine, and consisting of the colon (↓) and rectum (↓). It has the appearance of being divided into small bags. Its function is to absorb (p.163) water and mineral salts from the material passing through it, leaving faeces (p.169).

caecum (*n*) (*caeca*) a branch of the gut of an animal, closed at one end. Some vertebrates (p.148) have one or two caeca at the join of the small and large intestines; herbivorous (p.235) animals have a very large caecum which helps in the digestion of cellulose (p.138); carnivorous (p.235) animals have a very small caecum. In human beings, the caecum is a small bag at the start of the large intestine. **caecal** (*adj*).

vermiform appendix a small hollow finger-like structure leading from the caecum of some mammals; it contains lymphoid (p.182) tissue.

colon (*n*) the main part of the large intestine, concerned with the absorption of water and mineral salts. **colonic** (*adj*).

rectum (*n*) the end part of the large intestine; it has an opening at the end, either a cloaca (↓) or an anus (↓). Its function is to store faeces (p.169). **rectal** (*adj*).

anus (*n*) (*ani*) the opening at the end of the alimentary canal of mammals; through it passes the remains of undigested (p.166) food, i.e. faeces. **anal** (*adj*).

cloaca (*n*) the end of the alimentary canal of vertebrates, except mammals; tubes from the kidneys (p.185) and reproductive system (p.216) also enter into it. It has an opening to the exterior. Some invertebrates also have a cloaca. **cloacal** (*adj*).

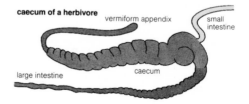

caecum of a herbivore

vermiform appendix

small intestine

large intestine

caecum

liver (*n*) a large gland (p.162) in many animals.
In vertebrates it has many functions including
(a) secretion of bile (p.168); (b) storage of
glycogen (p.174); (c) deamination (p.186).

gall bladder a small vessel, in or near the liver;
it stores bile (p.168) between meals. Bile leaves
by the bile duct to enter the duodenum (p.164).

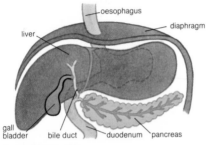

liver and gall bladder

bile duct a tube from the liver to the duodenum
(p.164); it conducts bile.

hepatic (*adj*) concerned with the liver, e.g. the
hepatic artery takes blood to the liver.

pancreas (*n*) a gland in all vertebrates (p.148)
except a few fishes; it produces an alkaline
digestive (↓) juice, called pancreatic juice.
Special groups of cells in the pancreas, called
islets of Langerhans, produce insulin (p.209).
The pancreatic duct, leads from the pancreas
into the duodenum (p.164). **pancreatic** (*adj*).

digestion (*n*) the chemical decomposition (p.95)
of food into substances which the body of an
animal can absorb (p.163). The chemical change
is brought about by digestive juices and, in most
animals, takes place in the gut. **digestive** (*adj*),
digestible (*adj*), **digest** (*v*).

juice (*n*) a liquid secretion (p.162) produced by
animals for the purpose of digesting food.

saliva (*n*) a liquid secretion (p.162) produced by
glands in the mouth. In land animals, saliva
contains mucus (p.190), and provides lubrication
(p.22) for food in the alimentary canal (p.163).
salivary (*adj*).

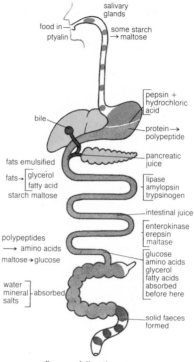

salivary glands

food in
ptyalin

some starch → maltose

pepsin + hydrochloric acid

protein → polypeptide

bile

pancreatic juice

fats emulsified

fats → glycerol
fatty acid

starch maltose

lipase
amylopsin
trypsinogen

intestinal juice

enterokinase
erepsin
maltase

polypeptides → amino acids

maltose → glucose

glucose
amino acids
glycerol
fatty acids
absorbed
before here

water
mineral salts

absorbed

solid faeces formed

diagram of digestive processes

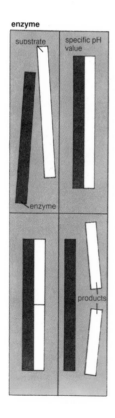

enzyme

substrate | specific pH value

enzyme

products

enzyme (*n*) any one of a large number of organic substances produced by all cells; an enzyme acts as a catalyst (p.99) on the chemical reactions which take place in all organisms (p.147). Most enzymes act on one substrate (p.146) only, so an organism produces a large number of enzymes. An enzyme is easily destroyed by heat, and by many chemical substances; it needs certain conditions, particularly a suitable pH value (p.116) before it will act. Most enzymes work inside cells; digestive (↑) enzymes are secreted into the gut and act on food there. **enzymatic** (*adj*).

ptyalin (*n*) an enzyme (p.167) present in the saliva (p.166) of some mammals, including humans. It acts on cooked starch (p.174).

gastric (*adj*) concerned with the stomach, e.g. the walls of the stomach produce gastric juice (p.166).

pepsin (*n*) an enzyme (p.167) which decomposes proteins. It is secreted in gastric juice (↑) and needs the presence of acid in order to work. Gastric juice contains hydrochloric acid to provide an acidic solution for pepsin.

bile (*n*) a green, alkaline liquid produced by the liver of vertebrates (p.148); it passes through the bile-duct from the liver to the duodenum (p.164). The function of bile is to form an emulsion (p.111) with fat, so that enzymes (p.167) can act readily on fatty foods. **biliary** (*adj*), **bilious** (*adj*).

lipase (*n*) an enzyme (p.167) which decomposes fats (p.175) into alcohols (p.132) and organic (p.131) acids. It is produced by the pancreas (p.166) and is one of the enzymes in pancreatic juice. Lipase acts only in an alkaline solution; it is passed into the duodenum, where bile forms an alkaline solution.

invertase (*n*) an enzyme (p.167) which decomposes sucrose (p.174) into glucose (p.174) and fructose (p.174). Both plants and animals produce invertase. In human beings, invertase is produced in intestinal juice (↓).

amylases (*n.pl.*) a group of enzymes which decompose starch or glycogen (p.174); the products include maltose and glucose (p.174). Ptyalin is salivary (p.166) amylase.

amylopsin (*n*) amylase (↑) produced by the pancreas (p.166) and present in pancreatic juice (↓).

trypsinogen (*n*) an inactive form of the enzyme trypsin (↓); it is made active by an enzyme, enterokinase (↓).

trypsin (*n*) an enzyme (p.167) which further decomposes proteins after the action of pepsin (↑). It is produced from trypsinogen (↑) by the action of enterokinase (↓). Trypsin acts only in an alkaline solution.

enterokinase (*n*) an enzyme (p.167) produced by glands in the wall of the small intestine (p.164). It acts on trypsinogen (↑).

lipase

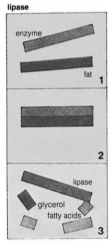

erepsin (*n*) a mixture of enzymes (p.167) which further decompose proteins after pepsin (↑) and trypsin (↑); it completes the decomposition of protein (p.172) to amino acids (p.172). Erepsin is secreted by glands in the wall of the small intestine (p.164).

maltase (*n*) an enzyme (p.167) which decomposes maltose (p.174) into glucose (p.174). It is secreted by glands in the wall of the small intestine (p.164).

pancreatic juice a secretion (p.162) of the pancreas; it contains lipase (↑), amylopsin (↑) and trypsinogen (↑). It passes to the duodenum (p.164) through the pancreatic duct.

intestinal juice a secretion (p.162) of glands in the walls of the small intestine (p.164). It contains invertase (↑), maltase (↑), enterokinase (↑), erepsin (↑) and completes the digestion of starch to glucose (p.174) and protein (p.172) to amino acids (p.172).

bolus (*n*) a round ball of food mixed with saliva ready to be passed down the oesophagus (p.164).

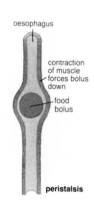

oesophagus

contraction of muscle forces bolus down

food bolus

peristalsis

wave of contraction passing down oesophagus

peristalsis (*n*) a wave-like motion of muscles along a tube-shaped vessel. The muscles contract (p.38), in turn, along the vessel, so that an object in the vessel is pushed along it. Peristalsis in the alimentary canal pushes its contents along and also mixes the food with digestive juices. **peristaltic** (*adj*).

faeces (*n.pl.*) the remains of undigested food. Water is absorbed from this material in the colon (p.165) to form faeces. The faeces are stored in the rectum (p.165) and passed out through the anus (p.165) from time to time.

defaecation (*n*) the action of passing out faeces through the anus (p.165). **defaecate** (*v*).

tooth (*n*) (*teeth*) (1) in vertebrates (p.148) a
  hard, usually sharp, structure in the mouth used
  for biting food or for attacking and seizing
  other animals. In mammals the teeth vary in
  shape according to their use, and are fixed in
  hollows, called sockets, in the jaw bone. (2) any
  small pointed structure sticking out, such as the
  teeth in a cog-wheel, *see diagram*. (3) a small
  pointed part of a leaf margin (p.158).

permanent teeth the second set of teeth which
  mammals (p.150) grow. The first set grows in
  young mammals; it contains fewer teeth than the
  second set, having no molars.

enamel (*n*) a hard substance forming the outer
  cover of teeth in mammals.

dentine (*n*) a hard bone-like substance which
  forms the main part of a tooth.

pulp-cavity a hollow inside the dentine of a
  tooth. It contains nerves, blood vessels, tissues
  and cells producing dentine. A narrow opening in
  the root of the tooth allows nerves and blood
  vessels to enter the pulp-cavity.

crown (*n*) the outside part of a tooth which is
  seen in the mouth; it is covered in enamel (↑).

incisor (*n*) a tooth with a cutting edge in the front
  of the mouth of a mammal; incisors are used for
  biting.

canine (*n*) a tooth which is pointed and sharp and
  grows behind the incisors. Herbivorous (p.235)
  animals have very small canine teeth, or none;
  carnivorous (p.235) animals have large canine
  teeth.

molar (*n*) a tooth with a flat surface, usually
  with two or three roots, growing at the back of the
  mouth. Molars are used for grinding food. There
  are no molars in the first set of teeth of a
  mammal.

premolar (*n*) a tooth similar to a molar, but with
  one or two roots; premolars grow between
  molars and canine teeth, and are present in the
  first set of teeth of a mammal.

dental (*adj*) concerned with teeth, e.g. dental
  decay. Compare **dentate** which describes an
  animal possessing teeth, or a leaf with small
  pointed parts. See tooth (↑).

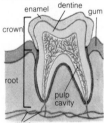

structure of a tooth

enamel dentine gum
crown
root
pulp
cavity

capillaries
and nerves

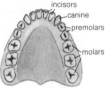

incisors
canine
premolars
molars

permanent teeth

teeth

a cog wheel

a simple food chain

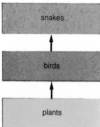

nutrition (*n*) the whole process of taking in food, digesting it, and using it to provide energy for living and materials for growth. The four stages of nutrition are: ingestion, digestion (p.166), absorption (p.163) and assimilation (p.163). **nutritious** (*adj*), **nutritional** (*adj*).

nutrient (*n*) a substance that can be used in the nutrition of an organism, e.g. (a) carbon dioxide is a nutrient for plants; (b) starch is a nutrient for human beings. **nutritive** (*adj*).

food chain a group of organisms (p.147) arranged in an order showing how each organism feeds on and obtains energy from the one before it, and is eaten and provides energy for the one after it. There are usually three or four organisms in a food chain. The first organism is a green plant, which obtains its food from inorganic compounds. The second organism is an animal which feeds on plants, a herbivorous animal. The third organism is an animal which eats herbivorous animals, a carnivorous animal. The fourth organism is a carnivorous animal which feeds on smaller carnivores. A typical food chain is: grass – insect – bird – snake.

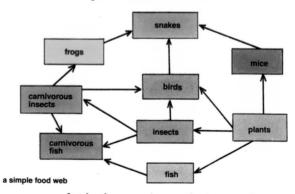

a simple food web

food web an organism may feed on more than one organism *before* it in a food chain, and that organism may provide food for more than one organism *after* it in the food chain. Many food chains can thus be joined together to form a food web.

metabolism (*n*) the chemical reactions (p.96) which take place in an organism (p.147) or part of an organism; the reactions are controlled by enzymes (p.167). Metabolism includes the decomposition of organic compounds, as in digestion, and the synthesis (p.136) of new compounds, as in the production of digestive juices. **metabolic** (*adj*).

amino acid an organic (p.131) acid with an amino group of atoms ($-NH_2$) in the molecule. The formula of an amino acid is shown in the diagram.

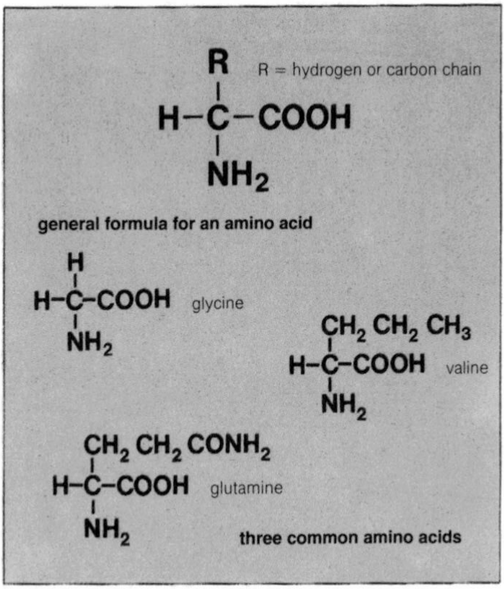

$$R$$

R = hydrogen or carbon chain

$$H-C-COOH$$
$$NH_2$$

general formula for an amino acid

$$H-C-COOH \quad glycine$$
$$NH_2$$

$$CH_2\ CH_2\ CH_3$$
$$H-C-COOH \quad valine$$
$$NH_2$$

$$CH_2\ CH_2\ CONH_2$$
$$H-C-COOH \quad glutamine$$
$$NH_2$$

three common amino acids

protein (*n*) a kind of compound formed by the chemical combination of many molecules of different amino acids. One protein molecule contains hundreds or thousands of amino acid molecules.

polypeptide (*n*) a compound formed by the chemical combination of molecules of amino acids, but containing fewer amino acids than a protein. Proteins are first decomposed by digestive juices to polypeptides. Polypeptides are then further decomposed to amino acids.

sources of carbohydrate in human diet

bread

rice

potato

carbohydrate (*n*) a compound consisting of carbon, hydrogen and oxygen with the atoms of hydrogen and oxygen in the same proportion as in water; the general formula of carbohydrate is $C_x(H_2O)_y$. Starch (p.174), cellulose (p.138) and all sugars (↓) are carbohydrates. Plants contain large amounts of carbohydrates, but animals contain much smaller amounts; in all organisms, carbohydrates are important in metabolism (p.172).

sugar (*n*) a crystalline (p.110) carbohydrate (↑) soluble in water and sweet to the taste. Plants produce various sugars in photosynthesis. A sugar is the simplest kind of carbohydrate.

monosaccharide (*n*) one of the simplest sugars (↑), containing either 5 or 6 carbon atoms. When a monosaccharide is decomposed, the products are no longer sugars, *see disaccharides*. The formula of a monosaccharide with 6 carbon atoms is $C_6H_{12}O_6$, and the atoms can be arranged in the molecule to form isomers (p.132). Glucose and fructose both have the same molecular formula $C_6H_{12}O_6$, but have different properties. Cell membranes are generally permeable to monosaccharides.

disaccharide (*n*) a sugar with a molecule made up of two monosaccharide (↑) molecules; the monosaccharide molecules can be the same, or different. Common disaccharides have 12 carbon atoms and a general formula of $C_{12}H_{22}O_{11}$. On hydrolysis (p.136) a disaccharide molecule is decomposed into two monosaccharide molecules. Isomers of disaccharides have different properties. Cell membranes are impermeable (p.156) to disaccharides and polysaccharides (↓).

polysaccharide (*n*) a carbohydrate with a molecule made up of many hundreds of monosaccharide molecules. The molecules are often fibrous. All plants store energy in polysaccharides; animals also use polysaccharides as an energy store, but to a lesser amount. On hydrolysis (p.136) polysaccharides are decomposed to disaccharides (↑) or monosaccharides (↑).

glucose (*n*) a monosaccharide (p.173) present in all plants and animals. It is produced in green plants by photosynthesis (p.159) from carbon dioxide and water, and stored as starch (↓). In animals, it is a last product of the digestion of carbohydrates and is stored as glycogen (↓). Its formula is $C_6H_{12}O_6$.

fructose (*n*) a monosaccharide (p.173) present in many plants. Its formula is $C_6H_{12}O_6$. Fructose can be absorbed by the gut (p.162).

maltose (*n*) a disaccharide, formula $C_{12}H_{22}O_{11}$, produced by the hydrolysis (p.136) of starch (↓). It is present in germinating (p.156) seeds, and is produced during the digestion (p.166) of starch. A molecule of maltose contains two molecules of glucose, chemically combined. Maltose, on hydrolysis, produces glucose.

sucrose (*n*) a disaccharide, formula $C_{12}H_{22}O_{11}$, present in many plants but not in animals. It is made from sugar cane. A molecule of sucrose contains one molecule of glucose (↑) combined with one molecule of fructose (↑). Sucrose is hydrolysed (p.136) to glucose and fructose (↑).

starch (*n*) a polysaccharide with a molecule formed from many molecules of glucose (↑) chemically combined. It is a white amorphous (p.110) substance, insoluble in water, and on hydrolysis (p.136) forms glucose. Plants store carbohydrates (p.173) as starch.

glycogen (*n*) a polysaccharide with a molecule formed from many molecules of glucose chemically combined; it is soluble in water. Animals and fungi (p.145) store carbohydrates (p.173) as glycogen. On hydrolysis (p.136) it forms glucose. In vertebrates it is present especially in the liver and in muscles.

glucose molecule

molecule of glycerol

$CH_3-(CH_2)_{14}-COOH$

molecule of palmitic acid

fat (*n*) true fat is an ester (p.132) of glycerol (↓) and one or more fatty acids (↓). On hydrolysis (p.136) it is decomposed to glycerol and a mixture of fatty acids. A fat is solid at room temperature, an oil (p.134) is liquid. Other substances which can be dissolved in hot alcohol are also known as fats, but they are not true fats. Butter is a true fat. **fatty** (*adj*).

glycerol (*n*) an alcohol (p.132) with three hydroxyl groups and formula $C_3H_6O_3$, *see diagram*; it is a sweet, sticky, odourless, colourless liquid.

fatty acid an organic acid with a straight hydrocarbon (p.131) chain. The fatty acids present in plants and animals usually have an even number of carbon atoms; they have a carboxyl group of atoms (−COOH) at the end of the chain. Common acids in plants are stearic acid ($C_{17}H_{35}COOH$); palmitic acid ($C_{15}H_{31}COOH$); oleic acid ($C_{17}H_{33}COOH$).

diet (*n*) the different kinds of food and the amount of each kind of food eaten by a person or by a group of people. It is often described by the name of the food which provides most of the energy for nutrition, e.g. a rice diet, a wheat diet.

Calorie[2] (*n*) a kilocalorie, i.e. 1 Calorie (capital C) = 1000 calories, a unit of energy used to measure the energy value of different kinds of foodstuffs

foodstuff (*n*) a chemical substance in food used by animals. Foodstuffs are carbohydrate (p.173), protein (p.172) and fat (↑); they provide the energy for living, and the substances needed for growth and for replacing worn out tissues.

calorific value the number of Calories provided by a known mass of a foodstuff.

energy value another name for calorific value.

vitamin (*n*) an organic (p.131) substance which an animal must obtain in its food in order to be healthy. Vitamins are only needed in small amounts, and part of the need may be synthesized (p.136) by an animal, although this does not often happen. Different animals require different vitamins. Every vitamin is available from another plant or animal and many can be synthesized in factories. **vitaminize** (*v*).

balanced diet a diet which supplies enough energy for a person to live, enough protein for him to grow new tissues with a balance between carbohydrate (p.173), protein (p.172) and fat (p.175). In addition a person needs enough vitamins (p.175), mineral salts and roughage (↓).

malnutrition (*n*) a condition in which the body of a person does not get a suitable diet. There can be a lack of protein or a lack of carbohydrate and fat or a lack of vitamins, or a lack of them all.

roughage (*n*) a part of food which cannot be digested, such as cellulose in man's diet (p.138). It helps peristalsis (p.169), and is a part of a balanced diet (p.175).

deficiency disease a disease caused by the lack of a vitamin or a mineral element such as iron or calcium, or an amino acid needed in the diet. The lack must be great to cause the disease. Deficiency diseases arise from malnutrition (p.176).

night blindness a deficiency disease (↑) in which a person sees by day but badly by night or in a bad light. It is caused by a lack of vitamin A.

xerophthalmia (*n*) a deficiency disease (↑) in which the cornea (p.204) of the eye becomes dry and the person finally becomes blind. It is caused by a lack of vitamin A, greater than the lack which causes night blindness (↑).

beri-beri (*n*) a deficiency disease (↑) in which the nerves fail to act, particularly in the legs, and a person is not able to walk. It is caused by a lack of vitamin B_1.

pellagra (*n*) a deficiency disease (↑) in which the skin becomes rough and brown, and a person's mind becomes ill. It is caused by a lack of vitamin B_7.

scurvy (*n*) a deficiency disease (↑) in which the teeth become loose, and a person can readily fall ill from other diseases. It is caused by a lack of vitamin C.

rickets (*n*) a deficiency disease (↑) in which the bones of young children become soft and their legs are deformed. It is caused by a lack of vitamin D.

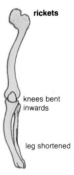

rickets

knees bent inwards

leg shortened

marasmus (*n*) a deficiency disease (↑) in children under the age of 5 years. The child is hungry, always crying, fails to grow, and its legs and body become thin. It is caused by a diet lacking in carbohydrates (p.173) and fats (p.175), i.e. a lack of foods which provide energy.

kwashiorkor (*n*) a deficiency disease (↑) in children under the age of 4 years. The child does not want to eat, has a swollen body, its hair becomes soft and changes colour, and it fails to grow. The disease is caused by a lack of protein (p.172) in the diet.

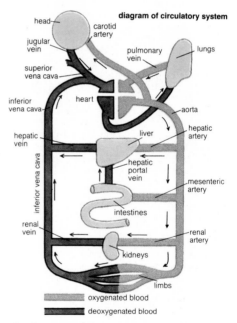

diagram of circulatory system

head — carotid artery
jugular vein
pulmonary vein — lungs
superior vena cava
inferior vena cava
heart
aorta
hepatic vein
liver
hepatic artery
inferior vena cava
hepatic portal vein
mesenteric artery
intestines
renal vein
renal artery
kidneys
limbs
▨ oxygenated blood
■ deoxygenated blood

circulatory system a system (p.162) of tubes and spaces in an animal, through which a liquid flows taking dissolved substances for the purpose of metabolism (p.172). It usually has an organ which pumps (p.33) the liquid round the system; in vertebrates the organ is the heart (p.181).

vessel (*n*) a tube for conducting (p.45) fluids in an organism, e.g. blood-vessels in vertebrates (p.148) conduct blood to all parts of the body.

artery (*n*) a blood vessel conducting (p.45) blood from the heart (p.181) to the tissues of the body. In vertebrates (p.148) an artery has a thick, muscular wall. One artery, the aorta (↓) leaves the heart (p.181) and branches again and again until smaller arteries reach every part of the body.

arteriole (*n*) a small artery (↑) with walls formed from smooth muscle. The autonomic nervous system (p.200) controls the muscle and thus controls the blood supply to the capillaries (↓).

vein (*n*) (1) a blood vessel conducting (p.48) blood from the tissues to the heart (p.181); its diameter is smaller than that of an artery. In vertebrates (p.148) a vein has a thin wall and contains valves (p.33) which allow blood to flow in one directory only. (2) a vascular bundle (p.160) in a leaf, *see venation* (p.158).

vascular system a system of vessels for the conduction (p.45) of fluids. In vertebrates (p.148) the fluid is usually blood and lymph (p.182), and the vascular system consists of the circulatory system (p.177) and the lymphatic system (p.182). In plants, the vascular system conducts dissolved mineral salts, water and synthesized (p.136) food materials.

aorta (*n*) in mammals (p.150) the main artery, which leaves the heart (p.181) and supplies blood to all parts of the body except the lungs. In human beings, blood passes through the aorta at a rate of 4 litres per minute.

vena cava in vertebrates, except fish, the main vein entering the heart (p.181) and bringing back blood from all parts of the body, except the lungs. The vein is divided in two, one vessel bringing blood from the head and arms, and the other vessel bringing blood from the rest of the body.

portal system a system of veins conducting blood from one capillary network (↓) to another capillary network, e.g. the hepatic portal system conducts blood from capillaries in the intestines (p.164) to capillaries in the liver (p.166).

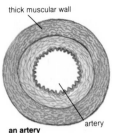

thick muscular wall

artery

an artery

thin muscular wall

vein

a vein

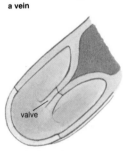

valve

a valve in a vein

a network of capillaries

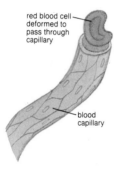

red blood cell
deformed to
pass through
capillary

blood
capillary

human red blood cells

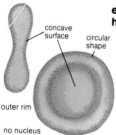

concave
surface

circular
shape

outer rim

no nucleus

capillary (*n*) a blood vessel with a very small diameter (about 10 μm) and very thin walls. The capillary walls are permeable (p.156) to water and ions of inorganic salts, together with dissolved oxygen, glucose, amino acids and carbon dioxide.

capillary network the arrangement of capillaries (↑) in tissues, in which the vessels branch and rejoin and so cover all the tissues.

sinus (*n*) a hollow space; it differs from a vessel in having a varying diameter. Blood sinuses are present in the circulatory system of some animals, especially invertebrates (p.148).

blood (*n*) in animals, a liquid contained in vessels or sinuses (↑), passed round a circulatory system (p.177) by a muscular pumping action. It contains dissolved products of digestion and excretions (p.186), and carries oxygen in blood cells (↓). In vertebrates, it consists of plasma (↓), red (↓) and white blood cells (p.180) and platelets (p.180).

plasma (*n*) a clear, almost colourless liquid left after blood cells have been taken out of blood of vertebrates (p.148).

red blood cell in vertebrates (p.148), a cell in the shape of a flat disc, containing haemoglobin (↓), which gives it a red colour. Red blood cells possess cytoplasm and a membrane, but, in mammals, have no nucleus; they have a life of about four months. They are elastic (p.27) and easily deformed to pass through capillaries; they have no power of motion. In human beings, there are about five million red blood cells in 1 mm³ of blood.

erythrocyte (*n*) another name for red blood cell.

haemoglobin (*n*) a red-coloured substance present in the red blood cells (↑) of vertebrates (p.148) and in the blood of some invertebrates, e.g. the earthworm. Each animal species has a different kind of haemoglobin; all combine readily with oxygen to form oxyhaemoglobin. Oxyhaemoglobin readily decomposes to set free oxygen, so haemoglobin acts as a carrier of oxygen from the lungs to the tissues. Haemoglobin is dark red and oxyhaemoglobin is bright red.

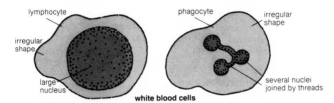

white blood cells

white blood cell in animal blood, a cell without
  colouring material; in vertebrates (p.148), it may
  be a phagocyte or a lymphocyte (p.182). In
  human beings, there are about 7000 white
  blood cells in 1 mm^3 of blood.
leucocyte (*n*) another name for white blood cell.
platelet (*n*) a small, flat part of a cell from bone
  marrow (p.192), present in the blood of
  mammals. In human beings, there are about
  250 000 platelets in 1 mm^3 of blood. Their
  function is to start the process of blood clotting
  (↓).
thrombin (*n*) an enzyme formed from a protein in
  the blood by the action of blood platelets, or
  injured tissues.
fibrin (*n*) an insoluble protein formed from a
  soluble protein, **fibrinogen**, by the action of the
  enzyme, thrombin (↑). Fibrin forms long fibres
  (p.195) in blood clotting (↓).
clot (*n*) a twisted net of fibrin (↑) fibres, which
  traps red blood cells to form a solid mass. A
  clot prevents blood escaping from a wound, and
  bacteria (p.145) from entering a wound; serum
  (↓) leaks out of a clot. **clot** (*v*).
serum (*n*) (*sera*) the liquid obtained from
  clotted (↑) blood; it is blood plasma without
  fibrin and the other substances needed to clot
  blood.
tissue fluid the liquid bathing all cells in an
  animal. It supplies the cells with glucose, amino
  acids, and fats in solution, i.e. the products of
  digestion. It takes away from cells carbon dioxide
  and any other unwanted products. Tissue fluid
  bathes capillaries and the capillary wall acts as a
  permeable membrane between blood and tissue
  fluid, allowing diffusion of dissolved substances.

homeostasis (*n*) the state of equilibrium (p.20) of the concentration (p.90) of dissolved substances in tissue fluid with the concentration of these substances in blood. The composition (p.95) of blood is controlled by various organs of the body so that it is kept constant (p.18) concerning: (a) osmotic pressure (p.139); (b) pH value (p.116); (c) concentration (p.90) of glucose; (d) concentration of amino acids. Homeostasis is the state of maintaining a constant composition of blood. Any change from these constant values causes damage to the body cells. **homeostatic** (*adj*).

heart (*n*) a hollow organ with muscular walls in the circulatory system (p.177) of an animal. Contractions (p.38) of the muscular walls pump blood round the system. In vertebrates (p.148), the heart is divided into auricles (p.182) and ventricles (p.182). Fishes have one auricle and one ventricle; amphibians have two auricles and one ventricle; birds and mammals have two auricles and two ventricles.

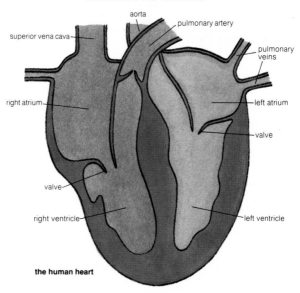

aorta
pulmonary artery
superior vena cava
pulmonary veins
right atrium
left atrium
valve
valve
right ventricle
left ventricle

the human heart

auricle (*n*) a chamber in the heart which receives blood from veins (p.178) and passes it to a ventricle (↓). It has muscular walls which are thinner than those of a ventricle. In vertebrates (p.148) with two auricles, one receives blood from the body and the other receives blood from the lungs. **auricular** (*adj*).

atrium (*n*) (*atria*) alternative name for auricle.

ventricle (*n*) a chamber in the heart which receives blood from an auricle (↑) and pumps blood, with its thick, strong, muscular walls, round the circulatory system (p.177). In vertebrates (p.148) with two ventricles, one pumps blood to the body and the other pumps blood to the lungs. **ventricular** (*adj*).

systole (*n*) the stage of contraction (p.38) of heart muscle in the action of the heart. Ventricles contract after auricles. **systolic** (*adj*).

diastole (*n*) the stage of relaxation (p.196) of heart muscle in the action of the heart. **diastolic** (*adj*).

lymph (*n*) a colourless liquid consisting of tissue fluid and white blood cells (p.180), mainly lymphocytes (↓). **lymphoid** (*adj*).

lymphatic system an arrangement of very small tubes, called lymph capillaries, take away tissue fluid (p.180). The lymph capillaries join to form larger tubes, called lymphatics; and the lymphatics join to form a lymph vessel which passes lymph into a main vein (p.178) near the heart. The lymph capillaries, lymphatics and lymph vessels form the lymphatic system. Lymph capillaries have walls which are more permeable (p.156) than blood capillaries, so even bacteria pass into the lymphatic system. Lymphatics have valves similar to those in veins.

lymphatic (*n*) a tube conducting lymph (↑); it collects lymph from lymph capillaries.

lymph node a small organ on a lymphatic (↑) consisting of lymphoid tissue. It produces lymphocytes (↓), removes bacteria from lymph, and filters out foreign bodies. Present in mammals and birds.

lymphocyte (*n*) a white blood cell (p.180) with a large nucleus and little cytoplasm; it produces antibodies (p.238).

the action of the heart

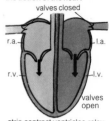

valves closed

r.a. — l.a.

r.v. — l.v.

valves open

atria contract ventricles relax

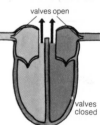

valves open

valves closed

ventricles contract atria relax

lymphatic system

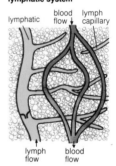

lymphatic   blood   lymph
             flow   capillary

lymph   blood
flow   flow

cells bathed in tissue fluid

lymph vessels

blood capillaries, oxygenated blood

blood capillaries, deoxygenated blood

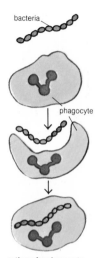

bacteria

phagocyte

action of a phagocyte

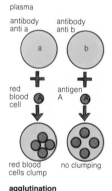

plasma

antibody anti a

antibody anti b

a

b

red blood cell

antigen A

red blood cells clump

no clumping

agglutination

phagocyte (*n*) a white blood cell (p.180), usually with several nuclei, with the power of amoeboid movement (p.144); it closes pseudopodia (p.144) round bacteria (p.145) and digests them. Phagocytes are important in defending animals against the attack of bacteria. **phagocytic** (*adj*).

phagocytosis (*n*) the action of phagocytes.

spleen (*n*) in vertebrates (p.148), an organ composed of lymphoid (↑) tissue, situated near the stomach. It is connected to the circulatory system (p.177). The spleen produces lymphocytes and takes worn out red blood cells (p.179) from the blood; it also stores red blood cells and supplies them to blood. **splenic** (*adj*).

septicaemia (*n*) a condition in which bacteria gets into the blood, and the blood is poisoned; the bacteria may come from a wound or may be bacteria causing a disease. Septicaemia causes a high temperature and red places on the skin.

blood group a group of people whose blood can be mixed without agglutination (↓). There are four blood groups, A; B; AB; O.

agglutination (*n*) sticking together. Red blood cells show agglutination, i.e. they stick together, when blood of different groups is mixed.

rhesus factor a substance present in the red blood cells of most people; such people are called rhesus positive (rh positive). The rest of the people do not have the substance present; they are rhesus negative. Rhesus negative persons do not possess an antibody (p.238) in their plasma (p.179) against the rhesus factor, but they can be given an antibody by a blood transfusion (↓). A rhesus negative woman bearing a rhesus positive child develops the antibody against the rhesus factor in the child. If the woman bears a second rhesus positive child, the unborn child can be damaged by the antibody in its mother's blood.

transfusion (*n*) the action of transferring blood from one person into another person. The blood given must be compatible (p.184). **transfuse** (*v*).

donor (*n*) a person who gives blood for a transfusion (↑) or who gives tissue or an organ to another person. **donate** (*v*).

recipient (*n*) a person who is given blood in a transfusion (p.183), or is given tissue or an organ in a medical operation. **receive** (*v*).

human blood groups

blood group of donor	blood group of recipient			
	O	A	B	AB universal recipient
O universal donor	✓	✓	✓	✓
A	✗	✓	✗	✓
B	✗	✗	✓	✓
AB	✗	✗	✗	✓

✓ compatible transfusions

✗ incompatible transfusions

compatible (*adj*) of blood, being able to be given in a transfusion. Red blood cells can have antigens (p.237) present called A and B. Red blood cells of group A have antigen A; of group B have antigen B; of group AB have antigens A and B; and of group O have no antigens. The plasma can have antibodies (p.238) present, called anti-A (or a) and anti-B (or b). Plasma of group A has antibody b; of group B has antibody a; of group AB has no antibodies; of group O has antibodies a and b. Plasma with antibody a agglutinates (p.183) red blood cells with antigen A; the cells stick together to form a clot (p.180). In a transfusion, donor (p.183) and recipient (↑) should be of the same blood group to prevent agglutination. Other blood groups can be used if the plasma of the recipient does not agglutinate the red blood cells of the donor. In these cases, the donor's blood is compatible with the recipient's. The table shows blood groups compatible for transfusion. **compatibility** (*n*).

incompatible (*adj*) of blood, not being able to be given in transfusion. If incompatible blood is given in a transfusion, the donor's blood forms a clot in the supply tube.

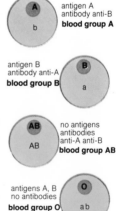

antigen A
antibody anti-B
blood group A

antigen B
antibody anti-A
blood group B

no antigens
antibodies
anti-A anti-B
blood group AB

antigens A, B
no antibodies
blood group O

human blood groups

universal donor a person of blood group O, who can donate (p.183) blood to people with blood of any group.

universal recipient a person of blood group AB who can receive (↑) blood in a transfusion (p.183) from people with blood of any group.

kidney (*n*) in vertebrates (p.148), one of two bean-shaped glands which control the amount of water in the body by taking water out of blood. A kidney also takes urea (p.186) and mineral salts out of the blood. It helps maintain homeostasis (p.181). It has an outer layer, the *cortex*, and an inner layer, the *medulla*.

diagram of a kidney

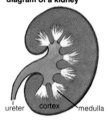

ureter   cortex   medulla

renal (*adj*) concerned with the kidneys, e.g. the renal artery supplies blood to the kidneys.

Bowman's capsule a cup-shaped structure, in a kidney, about 0.1 mm in diameter in man, with a knot of blood capillaries inside it. Urea (p.186), glucose (p.174), mineral salts, and water filter (p.91) through the walls of the capillaries and are taken away by a uriniferous tubule (↓).

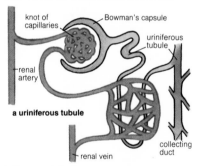

knot of capillaries   Bowman's capsule

uriniferous tubule

renal artery

a uriniferous tubule

collecting duct

renal vein

uriniferous tubule a narrow, coiled tube leading from the Bowman's capsule to collecting-ducts in the medulla of the kidney (p.185). The filtrate from the capsule passes along the tubule and glucose, mineral salts and water are absorbed (p.163) in it, and returned to the blood stream; the amounts absorbed keep the composition of the blood constant, *see* **homeostasis** (p.181).

uriniferous (*adj*) describes a tissue producing urine.

deamination (*n*) the taking away of an amino group from an amino acid (p.172), leaving an organic (p.131) acid. This action is done by the liver so that the concentration of amino acids in the blood is kept constant. The action forms ammonia (↓) from the amino group. **deaminate** (*v*).

ammonia (*n*) an inorganic compound with the formula of NH_3. It is poisonous to animals, so it is converted to urea (↓), a harmless compound, by the liver.

urea (*n*) an organic compound, soluble in water, with a formula of $CO(NH_2)$, formed from ammonia (↑) in animals; also present in plants.

urine (*n*) a liquid containing dissolved urea (↑) and some inorganic (p.116) salts; it is the liquid leaving a uriniferous tubule (p.185) and the kidney (p.185). **urinate** (*v*).

excrete (*v*) to send out waste products (i.e. those no longer needed) from the body, e.g. urea (↑) is a waste product of metabolism (p.172) as it is of no use to an animal; urea is excreted in urine. **excretion** (*n*), **excreta** (*n*).

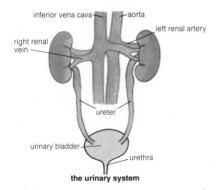

inferior vena cava — aorta

left renal artery

right renal vein

ureter

urinary bladder

urethra

the urinary system

urinary bladder a bag-like structure for storing urine until it is sent out of the body.

bladder (*n*) a shorter, less correct, name for urinary bladder.

ureter (*n*) a tube leading from a kidney (p.185) to the urinary bladder (↑).

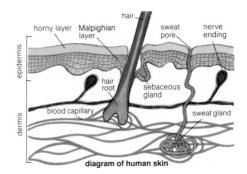

diagram of human skin

skin (*n*) the outer covering of an animal. In invertebrates (p.148) it consists of epithelium (p.192) on a membrane. In vertebrates (p.148) it consists of two layers (↓), epidermis (↓) and dermis (p.188).

layer (*n*) a flat, thin piece of material with similar flat pieces above and below it. The layers are thin in relation to their area; they may act as a cover to material beneath them. The strata (p.124) of rocks are in layers.

epidermis (*n*) the outside layer (↑) of cells of a plant or animal. In plants and invertebrates (p.148) the epidermis is only one cell thick. In vertebrates (p.148) the epidermis consists of a horny layer (↓) and a Malpighian layer (↓). **epidermal** (*adj*).

horny layer the outside layer (↑) of the epidermis (↑) of vertebrates (p.148) except fishes. It consists of dead cells which are slowly rubbed off and replaced by cells beneath. Its function is to prevent the entry of bacteria, and the loss of water from the body.

stratum corneum another name for horny layer (↑).
cornified layer another name for horny layer (↑).
Malpighian layer the layer (↑) of the epidermis between the horny layer (↑) and the dermis (p.188). It consists of actively dividing cells. (p.138). The outer cells of the Malpighian layer die and they replace the cells lost from the horny layer.

dermis (*n*) the inside layer (p.187) of cells in the skin of a vertebrate (p.148); it is much thicker than the epidermis (p.187). In this layer there are blood capillaries (p.179), nerve endings, hair roots, sebaceous glands (↓) and sweat glands (↓). The dermis provides the elastic strength of skin. **dermal** (*adj*).

sebaceous gland a gland (p.162) in the dermis (↑) usually opening on to a hair root. It produces sebum (↓).

sebum (*n*) an oily secretion (p.162) which keeps hair and skin soft and waterproof.

pore (*n*) a very small hole in a surface. The skin of mammals (p.150) has many pores in it. **porous** (*adj*).

sweat (*n*) a dilute solution of common salt, together with small amounts of other mineral salts, which is secreted (p.162) by sweat glands (↓). **sweat** (*v*).

sweat gland in mammals, a tube in the form of a knot, with blood capillaries around it; water and mineral salts are taken from the blood and sweat (↑) is formed. The sweat leaves the gland by a duct, a narrow tube, and passes through a pore (↑) to spread out over the skin. The production of sweat is controlled by the autonomic nervous system (p.200). The function of sweat glands and sweat is to control the temperature of warm-blooded (↓) animals, *see diagram* (p.187).

warm-blooded describes an animal which keeps its body at a constant (p.18) temperature, usually higher than that of the environment (p.226). Some heat is always lost by radiation (p.45); sweating, which cools the body by evaporation, provides the extra control to keep the temperature constant. Birds and mammals are warm-blooded.

homoiothermic (*adj*) warm-blooded (↑).

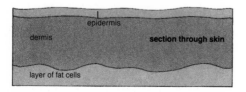

epidermis

dermis

section through skin

layer of fat cells

the thorax

thorax

gill cover

cold-blooded describes an animal whose body temperature is usually about the same as its environment (p.226). All animals except birds and mammals are cold-blooded. Animals living in water have a temperature very close to that of the water. Animals living on land have a temperature which may be, under some conditions, very different from that of the environment.

poikilothermic (*adj*) cold-blooded (↑).

thorax (*n*) (1) in vertebrates (p.148), except fish, the part of the body containing the lungs (↓) and heart; in mammals (p.150) it is separated from the abdomen (p.162) by a diaphragm (p.190), but not in other vertebrates. (2) in insects, the part of the body between the head and the abdomen; it carries the legs and wings.
thoracic (*adj*).

lung (*n*) an organ for breathing air in vertebrates (p.148); there are two lungs, one on each side of the heart. In the lung, oxygen is given to blood and carbon dioxide taken from blood, in capillaries.

gill (*n*) an organ of respiration (p.191) in most animals living in water; there are usually two gills, one on each side of the animal. Thin membranes (p.138) separate vessels conducting water and blood capillaries; through these membranes, dissolved oxygen from the water enters the blood, and carbon dioxide leaves the blood and dissolves in the water. The gills are inside the body of most organisms, but larvae (p.151) may have gills outside the body.

branchial (*adj*) describes anything to do with the gills, e.g. the branchial artery.

trachea (*n*) (*tracheae*) (1) in vertebrates (p.148) with lungs (↑), a tube leading from the mouth and nose, passing down the throat to the chest, where it branches into two bronchi (p.190); it conducts air down to the lungs. (2) in insects, a network of tubes reaching all parts of the body. Openings (spiracles) in the skin allow air to pass into the tracheae, which form the respiratory (p.191) system.

pulmonary (*adj*) concerned with the lungs, e.g. the pulmonary artery taking blood to the lungs.

bronchus (*n*) (*bronchi*) in vertebrates with lungs (p.189), a tube leading from the trachea (p.189) to each lung. A bronchus has plates of cartilage (p.192), as does the trachea, to prevent the tube closing. Both bronchus and trachea have glands secreting mucus (↓) and walls bearing cilia (p.144). The mucus removes dust, and the cilia beat to drive the dust back up to the mouth.

bronchiole (*n*) inside a lung a bronchus branches again and again, forming small tubes called bronchioles.

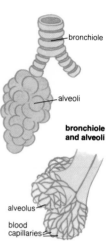

bronchiole

alveoli

bronchiole and alveoli

alveolus (*n*) (*alveoli*) a very small bag-like structure at the end of a bronchiole (↑). It has a network of blood capillaries on its surface, *see diagram*. Air inside an alveolus gives oxygen to blood and receives carbon dioxide from blood; these gases pass through the thin surface of the alveolus and the wall of a capillary. **alveolar** (*adj*).

alveolus

blood capillaries

pleura (*n*) (*pleurae*) a bag-like structure of very thin skin, which is the outer cover of a lung and also the inner coat of the space containing the lung. The skin is a membrane producing a watery liquid, which fills the space between the outer cover and the inner coat and prevents friction (p.22) between the two surfaces. **pleural** (*adj*).

diaphragm² (*n*) in mammals (p.150), a cup-shaped muscle between the thorax (p.189) and the abdomen (p.162). When the muscle contracts, the diaphragm becomes flatter, and air is drawn into the lungs.

inspire (*v*) to take air into lungs (p.189) or water into gills. Air is drawn into vertebrate (p.148) lungs by movement of the ribs. In mammals (p.150), the diaphragm is also used to inspire (inhale) air. **inspiration** (*n*).

expire (*v*) to push air out of the lungs (p.189) (exhale) or pass water out of the gills. Air is pushed out of vertebrate lungs by the relaxation (p.196) of the rib muscles and the diaphragm. **expiration** (*n*).

mucus (*n*) a sticky liquid secreted by special cells in membranes of vertebrates (p.148). Its function is lubrication (p.22). **mucous** (*adj*).

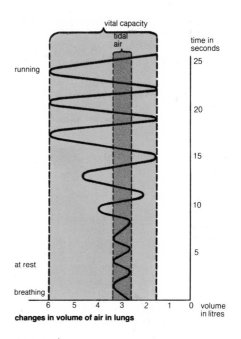

changes in volume of air in lungs

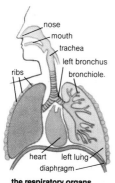

the respiratory organs

tidal air the volume of air taken in and sent out
of lungs (p.189) when breathing ordinarily. In
man, this is about 500 cm³.

vital capacity the volume of air taken in, after
having fully breathed out, until the lungs (p.189)
are completely full. In man, this is about 4000 cm³.

respiration (*n*) the whole process from the
inspiration (↑) of oxygen by lungs (p.189), gills
(p.189) or tracheae (p.189), through the use of
oxygen to supply energy in cells of the body, to
the expiration (↑) of carbon dioxide by lungs, gills
or tracheae. **External respiration** is the process
by which oxygen in the air is taken to cells of the
body, and carbon dioxide taken from the cells
and passed back to the air. **Internal** (or **tissue**)
respiration is the processes taking place in the cell
to use oxygen to provide energy for metabolism
(p.172). **respire** (*v*), **respiratory** (*adj*).

epithelium (*n*) (*epithelia*) a tissue which forms the surface or the outside of an organism, or is a covering for the inside of tubes and cavities in an organism. Examples of epithelium are the surface cells of the skin and the inside of blood vessels. Secretory (p.162) cells in glands are mostly epithelial tissue. Epithelium can also bear cilia (p.144), e.g. the inside of the walls of a bronchus are made of ciliated epithelium. **epithelial** (*adj*).

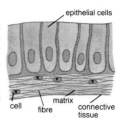

epithelial cells

cell fibre matrix connective tissue

epithelium and connective tissue

connective tissue a tissue in vertebrates, containing fibres or a matrix (↓); it provides support for other tissues and for organs, e.g. cartilage, bone, fatty tissue.

matrix (*n*) a solid material which is made by cells and secreted (p.162) round them, so that the cells are pushed apart leading to cells scattered here and there in the matrix.

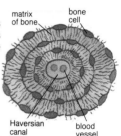

matrix of bone bone cell

Haversian canal blood vessel

structure of bone

cartilage (*n*) a connective tissue (↑) with cells scattered in an elastic (p.27) matrix containing polysaccharides (p.173) and many protein (p.172) fibres; the cells produce the matrix. Some animals, e.g. sharks, have cartilage instead of bone in their skeleton (↓); young children have cartilage which becomes changed to bone as they grow. **cartilaginous** (*adj*).

bone (*n*) a connective tissue of vertebrates (p.148) only; cells are scattered in a matrix (↑) which consists of fibres (p.195) and calcium salts. The inorganic (p.116) salts provide the hardness and the fibres the strength of bone. The bone cells are joined by small tubes carrying blood vessels and nerves. **bony** (*adj*).

Haversian canals small tubes in bone which carry blood vessels and nerves. Bone cells are arranged in circles round a Haversian canal.

marrow (*n*) soft material forming the inside of a bone. In long bones, *yellow marrow* fills the centre; it consists mainly of fat cells. The ends of the bone are filled with *red marrow*. Red blood cells (p.179) are produced in red marrow. Some white blood cells (p.180) are formed in yellow marrow.

ossification (*n*) the formation of bone, usually by the change of another tissue, such as cartilage, into bone. **ossify** (*v*).

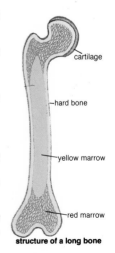

cartilage

hard bone

yellow marrow

red marrow

structure of a long bone

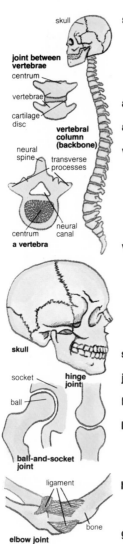

skull

joint between vertebrae

centrum

vertebrae

cartilage disc

vertebral column (backbone)

neural spine

transverse processes

centrum

neural canal

a vertebra

skull

socket

hinge joint

ball

ball-and-socket joint

ligament

bone

elbow joint

skeleton (*n*) a hard structure, either inside or outside an animal, which supports the animal's organs and tissues. Vertebrates (p.148) have an internal skeleton; some invertebrates (p.148), e.g. insects, have an external skeleton, while other invertebrates, e.g. earthworms, have no skeleton. **skeletal** (*adj*).

axial skeleton in vertebrates (p.148), the bones of the head and body.

appendicular skeleton the skeleton of the limbs of an animal.

vertebra (*n*) (*vertebrae*) a bone with a large central mass (the **centrum**), a hollow (the **neural canal**), and finger-like pieces of bone that stand out from the vertebra (**transverse processes** and **neural spine**), *see diagram*. The spinal cord (p.198) passes through the neural canal of a vertebra and is protected by the bone around it. **vertebral** (*adj*).

vertebral column the arrangement in a line of vertebrae (↑) joined by ligaments (↓) and separated by elastic cartilage (↑). It forms the main support of the body. The movement of any two vertebrae in relation to each other is small, but it is enough to allow the whole column to bend. Muscles are fixed to the finger-like pieces of bone which stand out from each vertebra, and control movement.

skull (*n*) the bones which protect the brain, together with those that form the face.

joint (*n*) the structure which joins two bones so that they can move in relation to one another.

ligament (*n*) a strong band of fibre (p.195) which keeps two bones of a joint held together.

ball-and-socket joint a joint in which the round end of one bone (the ball) fits into the hollow (the socket) of another bone; the bone can be turned in any direction by means of the ball turning in the socket, e.g. hip joint.

hinge joint a joint in which the round end of one bone turns on the flat surface of another bone, allowing the joint to bend in one plane only, e.g. knee joint.

gliding joint (*n*) a joint in which the surface of one bone moves over the surface of another bone.

movement (*n*) the action carried out when the bones in a joint are moved, e.g. waving the hands is a movement; it is not a motion, because the person does not change his place in space.

locomotion (*n*) the action or ability of moving from one place to another, i.e. being in motion without any outside help. In many animals, the movements of legs produce locomotion. **locomotor** (*adj*).

synovial capsule a bag-like membrane in a movable joint; it is fixed to the bones on either side of the joint, *see diagram*. The capsule is filled with a sticky liquid (synovial fluid) which lubricates (p.22) the joint.

pelvis (*n*) a bony structure in vertebrates (p.148) which provides strong support for the back legs or back fins. In mammals (p.150) it consists of several bones fixed together in the shape of a bowl. **pelvic** (*adj*).

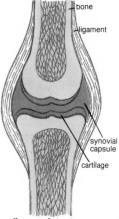

diagram of structure
of a joint

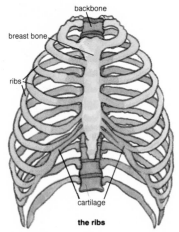

the ribs

rib (*n*) a narrow curved bone fixed to the vertebral column (p.193). The ribs form a protective structure round the thorax (p.189).

intercostal muscle a muscle (↓) between two ribs (↑). When the muscle contracts the ribs are raised and air is drawn into the lungs.

fibre of voluntary muscle

muscle (*n*) a tissue consisting of cells which have the power to contract. The contraction of muscles causes all movement of joints.

voluntary muscle in vertebrates (p.148), a tissue able to contract (p.38) rapidly, consisting of long muscle fibres (↓) with many nuclei (p.139) covered by a membrane. On receiving a stimulus (p.201) from the brain by way of a nerve, voluntary muscle contracts by becoming shorter and thicker. Muscle fibres are held together by fibres (↓) or connective tissue (p.192). Voluntary muscles are concerned with locomotion (↑) and movement by moving the bones of joints.

fibre (*n*) in animals, a thread-like structure of protein; it is very strong. In plants, a long thread-like structure of cellulose, e.g. cotton fibres. **fibrous** (*adj*).

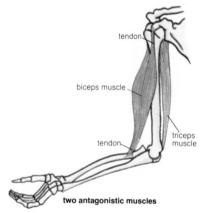

two antagonistic muscles

antagonism (*n*) of two muscles, producing movement in opposite directions, so that contraction of one muscle stretches the other. The only voluntary action of a muscle is contraction, so contractions of muscles in opposite directions are needed to control the movement at a joint. **antagonistic** (*adj*).

striped muscle another name for voluntary muscle, because it has stripes across it.

involuntary muscle in vertebrates (p.148), a tissue which contracts (p.38) slowly; it consists of long spindle-shaped cells bound together by connective tissue (p.192), and does not have the stripes of voluntary muscle (p.195). Involuntary muscle is usually found in sheets round hollow organs, e.g. intestines, blood vessels. This kind of muscle is controlled by the autonomic nervous system (p.200) and waves of contraction can pass along a muscle as in peristalsis (p.169).

smooth muscle another name for involuntary muscle.

cardiac muscle the muscle in the walls of the heart. It consists of a network of striped muscle fibres (p.195), but no membrane covers the fibres. The fibres contain separate cells, each with its nucleus (p.139). Cardiac muscle has the characteristics of both voluntary and involuntary muscle; its action is automatic (↓) and regular, faster than involuntary muscle but slower than voluntary muscle.

automatic (*adj*) describes an action or a process which, once started, continues to act by itself without being controlled by outside conditions, e.g. the action of heart muscle, the process of breathing.

relax (*v*) of muscles, to go from a state of acting to a state of rest. A muscle is either contracted or relaxed; in antagonistic (p.195) pairs of muscles, one contracts while the other relaxes. **relaxation** (*n*).

tendon (*n*) a band or string-like piece of connective tissue which attaches a muscle to a bone. A tendon consists of parallel fibres (p.195).

coordination (*n*) of muscles, the state of acting together to produce the same effect, e.g. when walking, the muscles of the legs act together to make each leg move in turn, and the muscles of the body keep it upright; all the muscles used in this process are working in coordination. **coordinate** (*v*).

flex² (*n*) to become bent; when a joint is flexed the angle between the two bones which form the joint becomes smaller. The opposite action is to extend a joint.

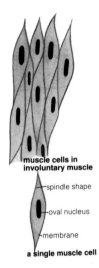

muscle cells in involuntary muscle

spindle shape

oval nucleus

membrane

a single muscle cell

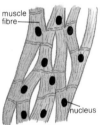

muscle fibre

nucleus

cardiac muscle

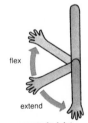

flex

extend

movement of a joint

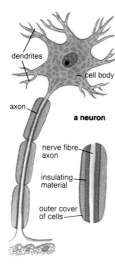

a neuron

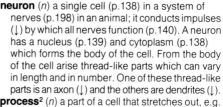

dendrites

cell body

axon

nerve fibre
axon

insulating
material

outer cover
of cells

neuron (*n*) a single cell (p.138) in a system of
nerves (p.198) in an animal; it conducts impulses
(↓) by which all nerves function (p.140). A neuron
has a nucleus (p.139) and cytoplasm (p.138)
which forms the body of the cell. From the body
of the cell arise thread-like parts which can vary
in length and in number. One of these thread-like
parts is an axon (↓) and the others are dendrites (↓).

process[2] (*n*) a part of a cell that stretches out, e.g.
an axon (↓).

impulse (*n*) a change, partly chemical and partly
physical, in a neuron, which is conducted by
axons, dendrites and cell bodies; it acts as a
message sent through a neuron (↑). An
impulse is like a travelling wave and it acts along
a length of between 2 and 5 cm of a nerve-fibre
(↓); it moves with a speed between 1 and
100 cm/sec, depending on the kind of nerve
(p.198) and the species (p.148) of animal. The
energy for the impulse is provided by the neuron,
so the impulse leaves a neuron with the same
energy as it entered the neuron.

axon (*n*) a long, thread-like part of a neuron (↑),
see diagram. It conducts impulses (↑) away from
the cell body of a neuron. A neuron has only one
axon.

dendrite (*n*) a thread-like part of a neuron (↑),
usually short in length, with branches at its end,
see diagram. It conducts impulses (↑) towards
the cell body of a neuron. A neuron can have
one or several dendrites.

synapse (*n*) the meeting place of dendrites (↑) and
axons (↑). The end of an axon touches the end
of a dendrite, and an impulse (↑) jumps from
the axon to the dendrite. An axon can have
several synapses, each with dendrites (↑) of
different neurons (↑). A dendrite can have
synapses with several axons. Through
synapses, an impulse can travel along many
paths through different neurons. **synaptic** (*adj*).

nerve-fibre an axon or a dendrite with a cover of
a fatty material for insulation (p.74). The diameter
of a nerve-fibre in vertebrates (p.148) is between
1 and 20 μm. Some very large nerve-fibres in
invertebrates are up to 1 mm in diameter.

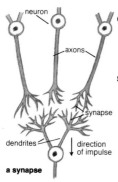

a synapse

neuron

axons

synapse

dendrites

direction
of impulse

nerve (*n*) a bundle of nerve-fibres (p.197) supported by connective tissue (p.192) with blood
vessels (p.178). Each nerve-fibre conducts
impulses independently of the others. All nerves
are motor (p.201) or sensory (p.200) or both in
function (p.140). A nerve can be a millimetre long
or as long as the animal. **nervous** (*adj*).

ganglion (*n*) (*ganglia*) a solid mass of cell
bodies of neurons (p.197); nerves enter and
leave it. **ganglionic** (*adj*).

spinal cord in vertebrates (p.148) a cylindrical
mass of nervous tissue containing cell bodies of
neurons, nerve-fibres and synapses; it is divided
into white matter (↓) and grey matter (↓). The
spinal cord runs through and is protected by the
vertebrae (p.193); a pair of spinal nerves leave
the spinal cord through holes in each vertebra.
Very simple coordination (p.196) by reflex arcs
(p.202) is effected in the spinal cord.

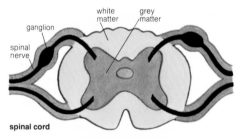

spinal cord

cerebrospinal fluid a liquid which fills the hole in
the middle of the spinal cord and fills spaces in
the brain. It is a solution of glucose and mineral
salts, and contains a few blood cells but no
protein in solution.

white matter a kind of nervous tissue consisting
of nerve fibres (p.197) and connective tissue
(p.192); it is the outer part of the spinal cord (↑)
and the inner part of the brain (↓).

grey matter a kind of nervous tissue consisting
mainly of cell bodies of neurons (p.197) with
dendrites and synapses and blood vessels; it is
the inner part of the spinal cord (↑) and the outer
part of the brain.

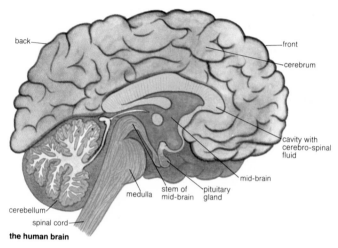

back — — front
— cerebrum

cavity with
cerebro-spinal
fluid

mid-brain

medulla   stem of   pituitary
           mid-brain gland

cerebellum —

spinal cord —

the human brain

brain (*n*) a large mass of ganglia (↑) protected in
vertebrates (p.148) by the bones of the skull. It
coordinates (p.196) actions of the whole body
particularly those in response to stimuli (p.201).

hind-brain this part of the vertebrate (p.148) brain
has two divisions, the medulla (↓) and the
cerebellum (↓); it joins the spinal cord (↑).

mid-brain a small part joining the hind-brain to
the fore-brain.

fore-brain the part containing the cerebrum
(p.200); in man it is the largest part of the brain.

medulla (*n*) part of the brain (↑) joined to the
spinal cord. It is concerned with (a) the
coordination (p.196) of impulses (p.197) from
hearing, tasting and touching; (b) the control of
respiratory (p.191) movements, cardiac muscle
(p.196), and blood vessels.

medulla oblongata fuller, more correct name for
medulla (↑).

cerebellum (*n*) a part of the brain (↑) growing out
from the stem of the hind-brain. It is concerned
with the coordination (p.196) of all muscular
movement, particularly the muscles fixed to
bones, e.g. coordination as needed in
locomotion (p.194).

cerebrum (*n*) in vertebrates (p.148) this consists of a pair of outgrowths from the front of the fore-brain. It is concerned with stimuli (↓) from receptors (↓) and sends impulses (p.197) by motor (↓) nerves to cause action in voluntary muscle (p.195) in answer to stimuli.

cerebral (*adj*) describes any tissue or effect which is to do with the cerebrum (↑).

meninges (*n.pl.*) in vertebrates (p.148), the three membranes (p.138) covering the spinal cord and the brain.

dura mater strong connective tissue (p.192) containing blood vessels; it is the outermost membrane (p.138) of the meninges (↑).

arachnoid (*n*) a membrane (p.138) between the dura mater (↑) and the pia mater (↓). It is separated from the pia mater by a space filled with cerebrospinal fluid (p.198).

pia mater thin membrane (p.138) containing many blood vessels; the innermost membrane of the meninges (↑).

central nervous system the nervous tissue which coordinates (p.196) all the activities of an animal. In vertebrates (p.148) it consists of the brain (p.199) and spinal cord (p.198). In many invertebrates it consist of a few large nerve-fibres (p.197) joined to several ganglia (p.198).

peripheral nervous system the nervous tissue in an animal, other than the central nervous system (↑). It consists of nerves and nerve fibres (p.197) which leave the central nervous system and branch to every part of the body.

autonomic nervous system in vertebrates, (p.148) motor (↓) nerves supplying involuntary muscle (p.196), the heart (p.181) and glands (p.162) together with sensory (↓) nerves from receptors (↓) inside the body. A lot of the co-ordination (p.196) between these nerves takes place in the spinal cord (p.198) and the medulla (p.199). The actions caused by the nerves are mainly automatic and take place independently of outside stimuli (↓).

sensory (*adj*) of nerves, concerned with receptors (↓). A stimulus (↓) causes an impulse (p.197) to travel along a sensory nerve.

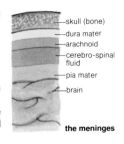

skull (bone)
dura mater
arachnoid
cerebro-spinal fluid
pia mater
brain

the meninges

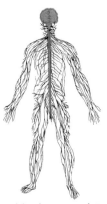

peripheral nervous system

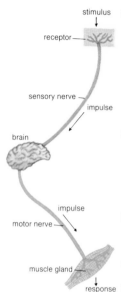

stimulus

receptor

sensory nerve

impulse

brain

impulse

motor nerve

muscle gland

response

irritability in animals

stimulus (*n*) (*stimuli*) any change in the outside
conditions of an organism which produces an
effect in the organism, but does not provide
energy for the effect. In animals, any change in
the outside conditions which is the cause of an
impulse in its nervous system, e.g. (a) the
direction of light is a stimulus for many plants;
(b) the smell of food is a stimulus for the salivary
glands (p.166); (c) seeing a snake is a stimulus
for a bird. **stimulate** (*v*), **stimulation** (*n*).

response (*n*) the change in an organism, or the
effect on an organism, produced by a stimulus
(↑), e.g. (a) under the stimulus of the direction of
light, the response of a plant is to grow towards
the light (see *phototropism*, p.161); (b) the
stimulus of seeing a snake produces a response
in a bird of flight. **respond** (*v*).

irritability (*n*) the ability to make a response (↑) to
a stimulus (↑). All organisms have irritability
while alive; it is a characteristic of life.
irritate (*v*).

irritable (*adj*) describes an organism which makes
a response (↑) to a stimulus (↑).

motor[2] (*adj*) of nerves, concerned with producing
action in muscles, glands (p.162), cilia (p.144),
e.g. (a) a motor nerve stimulates (↑) a muscle to
contract and flex (p.196) a joint; (b) a motor
nerve stimulates glands in the stomach wall to
secrete gastric juice (p.166).

receptor (*n*) an organ containing nervous tissue,
which responds (↑) to a particular stimulus (↑).
Different kinds of receptors respond to different
kinds of stimuli. When a stimulus is strong
enough to have an effect on a receptor, the
receptor sends an impulse (p.197) along the
sensory (↑) nerve connected to the receptor,
e.g. the eye is a receptor for the stimulus of
light and it sends impulses which travel along an
optic nerve to the brain. **receptive** (*adj*),
receptivity (*n*), **receive** (*v*).

sense-organ another name for receptor.

sense (*n*) the ability to receive stimuli (↑) from
the environment (p.226). The senses are: seeing;
hearing; smelling; tasting; touching; feeling
pain, heat, cold. **sense** (*v*), **sensory** (*adj*).

tactile (*adj*) concerned with touch, e.g. a tactile
corpuscle is an end-organ (↓) of touch.

end bulb (*n*) a small end organ (↓) which is a
receptor (p.201) in the skin.

end-organ a small organ containing one or a few
cells; it is connected to the central nervous
system (p.200) by a nerve-fibre (p.197). It may
be a receptor (p.201) or it may change an
impulse (p.197) into a stimulus (p.201) for a
muscle or gland.

taste-bud the receptor (p.201) of taste, present in
four groups on the surface of the tongue. Each
group responds to one of the four tastes: bitter,
sweet, sour, salty. When food is tasted, the
flavour is sensed by the receptors for smell in the
nose.

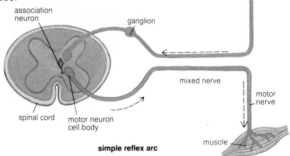

simple reflex arc

reflex (*n*) a very simple kind of behaviour seen
in all animals with a nervous system. A particular
stimulus (p.201) always causes the same
response (p.201) with no delay, e.g. an object
aimed at the eye causes the reflex action of the
eyelid closing.

reflex arc the path followed by an impulse (p.197)
from receptor to end-organ in a reflex, *see
diagram*. The path is: receptor, sensory (p.200)
nerve, synapse, association neuron (↓),
synapse, motor (p.201) nerve, end-organ.

association neuron a neuron in the spinal cord
(p.198) which, by synapses, joins a sensory
(p.200) nerve to a motor (p.201) nerve. It also
has synapses to pass an impulse (p.197) to other
parts of the spinal cord and to the brain (p.199).

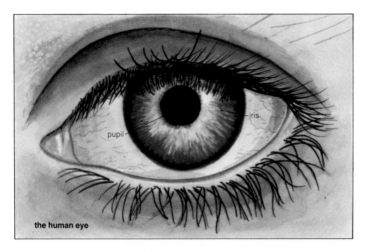

the human eye

eye (*n*) in animals, the receptor for light. The
structure of an eye varies greatly from a simple
organ – that responds (p.201) only to light, such
as seen in some protozoa (p.143) – to the
human eye.

eyeball (*n*) in vertebrates, the ball-shaped eye
containing nervous tissue stimulated by light.

iris (*n*) the coloured circular muscle in the front
of the eye; it contains a central opening, the pupil
(↓). It controls the amount of light entering the
eye.

pupil (*n*) the opening in the iris (↑) through which
light enters the eye. The size of the pupil is
controlled by the iris. The amount of light is
a stimulus (p.201) for a reflex (↑) controlling the
response (p.201) of the iris.

lachrymal gland in vertebrates (p.148), other than
fish, a gland which secretes (p.162) a slightly
antiseptic (p.240) liquid (tears) which keep the
cornea (p.204) wet.

sclerotic coat in vertebrates (p.148), the strong
outer layer of the wall of the eyeball, consisting
of fibrous or cartilaginous (p.192) connective
tissue (p.192). It gives shape to the eyeball and
protects the inner parts.

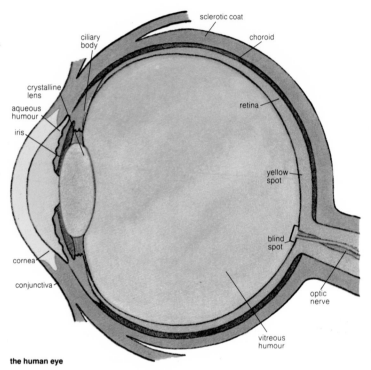

the human eye

cornea (*n*) in vertebrates (p.148), the transparent
(p.52) covering over the iris (p.203) and the
crystalline lens (p.59); it is part of the sclerotic
coat. The cornea in land vertebrates produces
most of the refraction (p.57) needed to focus
(p.58) an image (p.56) on the retina (↓). The
crystalline lens completes the refraction.
corneal (*adj*).

conjunctiva (*n*) in vertebrates (p.148), a layer of
epidermis (p.187) which covers the cornea (↑)
and the part of the sclerotic coat (p.203) seen as
the white part of the eye. The conjunctiva
secretes (p.162) mucus (p.190); it is transparent
(p.52) and prevents bacteria from entering the
eyeball.

choroid coat the middle layer of the wall of the
eyeball in some vertebrates (p.148). It is dark
coloured and absorbs (p.163) light. The choroid
coat carries many blood vessels to supply the
retina (↓) and other parts of the eye.

retina (*n*) in vertebrates (p.148), the inner layer of
the wall of the eyeball. It contains nervous tissue
which is stimulated (p.201) by light, to send
impulses by the optic (↓) nerve to the brain.

optic (*adj*) concerned with the eye, e.g. the optic
nerve, which goes from the eye to the brain.

blind spot the place where the optic nerve enters
the eye of vertebrates (p.148). There is no retina
at this point, so light does not stimulate nervous
impulses at the blind spot.

yellow spot an area of the retina where sight is
clearest, present in man and some apes. When
a person looks at an object, the image (p.56)
is focused (p.58) on the yellow spot.

ciliary body a ring of muscle and supporting
tissue at the edge of the choroid coat (↑). It
contains the ciliary muscle, which alters the
shape of the crystalline lens (p.59) and gives the
lens its power of accommodation (p.59). The iris
is fixed to the ciliary body. The ciliary body
secretes (p.162) aqueous humour (↓).

aqueous humour a watery liquid which fills the
space between the cornea (↑) and the crystalline
lens (p.59). It helps to keep the shape of the
eyeball.

vitreous humour a jelly-like material which fills
the eyeball from the crystalline lens (p.59) to the
retina. It keeps the shape of the eyeball and also
helps the lens by further refraction (p.57) of light.

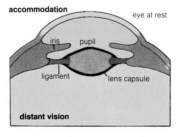

accommodation

eye at rest

iris   pupil

ligament   lens capsule

distant vision

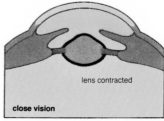

lens contracted

close vision

outer ear in mammals (p.150), a short tube leading
to the ear-drum from the pinna (↓). In birds,
there is no pinna. In amphibians (p.149) and most
reptiles (p.150) there is no outer ear; the ear-
drum (↓), if present, is in the skin. Fish do not
possess an ear.

pinna (*n*) in mammals, structure of skin and
cartilage fixed to the head; the only part of the
ear that can be seen. It helps to collect sound
vibrations (p.64).

ear-drum a thin membrane (p.138) stretched over
and closing the tube of the outer ear. It vibrates
(p.64) to sound, and passes the vibrations to the
ossicles (↓).

middle ear in vertebrates (p.148), other than fish,
a narrow air-filled space containing the ossicles
(↓).

ossicle (*n*) a small bone in the middle ear. There
are three ossicles in mammals and they pass
vibrations (p.64) from the ear-drum (↑) to the
oval window (↓). In birds, reptiles (p.150) and
many amphibians (p.149) there is only one
ossicle.

Eustachian tube a tube leading from the middle
ear (↑) to the throat in land vertebrates (p.148).
Its function is to make the air pressures equal on
each side of the ear-drum (↑).

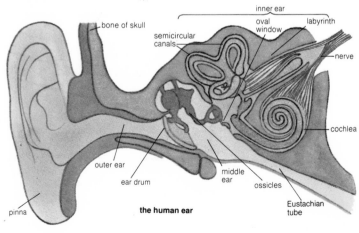

the human ear

inner ear in land vertebrates (p.148), the receptor (p.201) for sound, and the sense organ of balance (p.18); it contains the labyrinth (↓).

oval window a thin membrane (p.138) between the middle ear and the inner ear (↑). The ossicles pass vibrations (p.64) to the oval window. The oval window passes vibrations to the labyrinth (↓).

labyrinth (*n*) a system (p.162) of tubes and hollows in the inner ear (↑). The bony labyrinth is in the side of the skull (p.193). The membranous (p.138) labyrinth consists of a membrane filled with liquid and fits inside the bony labyrinth.

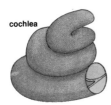

cochlea

cochlea (*n*) a tube in the shape of a spiral (p.219), part of the labyrinth (↑). The liquid in the labyrinth conducts sound vibrations (p.64) to the cochlea, and it responds to the pitch (p.65) of the sound. The cochlea senses both loudness and pitch, and sends impulses to the brain by a nerve.

organ of Corti (*n*) the part of the ear that hears sounds; it is in the cochlea. Sound vibrations enter the pinna (↑) and are passed by the eardrum and ossicles to the oval window (↑) which then passes sounds to the fluid in the cochlea. Each part of the organ of Corti responds to a different pitch (p.65). Hair cells on the organ send impulses through nerve fibres to the auditory nerve, connected to the brain.

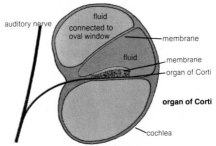

auditory nerve

fluid connected to oval window

membrane

fluid

membrane

organ of Corti

organ of Corti

cochlea

semicircular canals in vertebrates (p.148), three tubes, semicircular in shape; placed at right angles to each other and fixed to the labyrinth (↑). They are concerned with the sense of balance, and detect any turning movement of the animal.

endocrine gland in vertebrates (p.148), a gland (p.162) which secretes (p.162) a hormone (↓). The gland has no duct (p.162), the hormone diffuses (p.27) into the blood stream from the gland.

ductless gland another name for endocrine gland (↑).

hormone (*n*) an organic (p.131) substance produced in very small amounts by endocrine glands (↑) in one part of an animal and carried by the blood stream to another part where it has an important effect. Plants also produce hormones in very small amounts and hormone movement in the plant is controlled by the plant cells.

adrenalin (*n*) a hormone (↑) secreted by the *adrenal glands* of vertebrates (p.148). Some invertebrates (p.148) also secrete adrenalin. Adrenalin has the following effects: (1) increases the rate of heart beat; (2) widens the blood vessels of muscles, brain and heart; (3) narrows the blood vessels of skin and viscera (p.162); (4) makes the pupil (p.203) of the eye larger; (5) makes hair stand up; (6) increases the amount of sweat (p.188); (7) speeds the change of glycogen (p.174) to glucose (p.174). The stimulus (p.201) for the nervous system is fear, anger or pain. The autonomic nervous system (p.200) then stimulates the adrenal gland, and the body is made ready to escape from danger.

thyroid gland in vertebrates (p.148) an endocrine gland (↑) secreting (p.162) a hormone, **thyroxin**. Thyroxin is a substance containing iodine; it controls the rate of metabolism (p.172) and so controls growth and the production of heat in the body. Lack of iodine in the diet causes the thyroid gland to become very large, a condition known as **goitre**. The thyroid gland is controlled by a hormone from the pituitary gland (↓).

pituitary gland in vertebrates (p.148) an endocrine gland (↑) beneath the floor of the brain. It secretes (p.162) a number of hormones (↑) which control the action of other endocrine glands. The pituitary gland is the most important endocrine gland; it is under the direct control of the central nervous system (p.200).

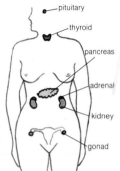

some endocrine glands in a human

insulin (*n*) a hormone (↑) secreted (p.162) by special cells in the pancreas (p.166) of vertebrates (p.148). Secretion is stimulated (p.201) by a high concentration (p.90) of glucose (p.174) in the blood. Insulin changes glucose to glycogen (p.174) and controls the concentration of glucose in the blood. Lack of insulin causes the disease **diabetes**, in which glucose is present in urine; the kidneys excrete glucose to lower the concentration of glucose in blood. A complete lack of insulin causes death. Insulin and adrenalin (↑) are antagonistic (p.195) in their actions.

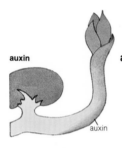

auxin

auxin (*n*) one of a group of plant hormones produced by actively dividing cells at the ends of stems and roots. An auxin increases the growth in length of a plant cell, thus causing the curving of a stem, or root, in phototropism (p.161) and geotropism (p.161). Auxins also control fruit growth, leaf fall, growth of buds (p.213) and other kinds of plant growth.

reproduction (*n*) of organisms (p.147), the process of producing young organisms which have the same kind of characteristics (p.147) as the parent organism. Reproduction can be either sexual or asexual (↓). **reproduce** (*v*), **reproductive** (*adj*).

sexual reproduction reproduction (↑) in which two organisms (p.147) take part, one called *male* and the other called *female*. Male and female organisms have different reproductive structures (p.144); a male gamete (p.210) is given by a male to a female to unite with a female gamete to produce a new organism.

asexual reproduction reproduction (↑) in which only one organism takes part; it is reproduction without gametes (p.210). In plants it is carried out by forming spores (p.146) or by vegetative reproduction (p.213). In animals it is carried out by binary fission (p.141) or by budding (p.213).

fertile (*adj*) of gametes (p.210) and organisms (p.147), capable of reproducing a young organism. **fertility** (*n*).

sterile (*adj*) (1) of organisms (p.147), not capable of sexual reproduction (↑). (2) free from bacteria, viruses, fungi and protozoa (p.143).

gamete (*n*) a reproductive (p.209) cell with a
   haploid (p.220) number of chromosomes (p.142)
   in its nucleus (p.139). In many organisms
   (p.147) the male and female gametes are
   different. The female gamete usually has a large
   cytoplasm (p.138) and is not capable of locomotion (p.194). The male gamete usually has very
   little cytoplasm and is capable of locomotion by
   means of a flagellum (p.144).

sex-cell another name for gamete.

gonad (*n*) in animals, the organ which produces
   gametes (↑). In some animals, gonads also
   produce hormones.

fusion (*n*) the process of two nuclei (p.139) from
   gametes (↑) joining and becoming one nucleus.

fertilization (*n*) the union of two gametes (↑),
   which takes place in two stages (p.152); (1) the
   fusion (↑) of the nuclei of two gametes; (2) the
   start of the development of a new organism.

external fertilization is the union of gametes
   outside the body of the parents. **fertilize** (*v*).

internal fertilization is the union of gametes
   inside the body of the female parent.

zygote (*n*) a fertilized female gamete (↑) before
   it starts to grow by binary fission (p.141). It is the
   first cell of the new individual in sexual reproduction.

offspring (*n*) (*offspring*) the young animals
   reproduced (p.209) by their parents.

generation (*n*) a single stage (p.152) in a family;
   the parents form one generation and their
   offspring (↑) the next generation.

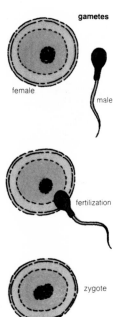

gametes

female

male

fertilization

zygote

sexual
fertilization

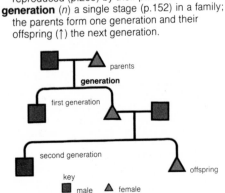

parents

generation

first generation

second generation

key
■ male ▲ female

offspring

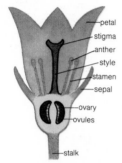

diagram of a flower

flower (*n*) a reproductive structure in monocotyledon (p.155) and dicotyledon (p.155) plants. The *essential* flower parts are concerned with reproduction, while the *accessory* flower parts are not directly concerned with reproduction.

stamen (*n*) the male essential flower part. It consists of a thin stalk (p.157), called a filament, bearing an anther (↓). **staminate** (*adj*).

anther (*n*) a structure at the end of the stamen (↑) stalk consisting of two parts, each containing pollen (↓) sacs. An anther produces pollen (↓).

filament^2 (*n*) a thin stalk (p.157) which supports an anther (↑) in a position to effect cross-pollination (↓).

pollen (*n*) a small grain which produces two male gametes. Some pollen grains are very light and easily blown by the wind; other pollen grains are heavier and sticky, and are carried by insects.

pollination (*n*) the process by which pollen grains are carried from anther (↑) to stigma (↓). This is done either by wind or by insects (p.151). The pollen is carried from the anther (↑) of one plant to the stigma (↓) of another plant in **cross-pollination**. In **self-pollination**, pollen is carried from anther to stigma of the same flower or to a stigma of another flower on the same plant.

carpel (*n*) a female essential part of a flower (↑). It consists of a stigma (↓), a style (↓) and an ovary (↓) containing ovules (↓). A flower may contain more than one carpel.

stigma (*n*) the surface of a carpel (↑) which is usually sticky; pollen grains stick to it and germinate (p.156).

style (*n*) a short stalk bearing a stigma (↑). After a pollen (↑) grain has germinated (p.156), a pollen tube grows down the style to an ovule (↓), and male gametes pass down the tube.

ovary^1 (*n*) a hollow space at the bottom of a carpel (↑) with a thick wall around it. The ovary contains one or more ovules (↓). It develops into a fruit (p.155).

ovule (*n*) in seed plants, a structure that contains a female gamete (p.210) which, after fertilization by a male gamete, forms an embryo. Each ovule is fixed to the ovary (↑) wall by a stalk.

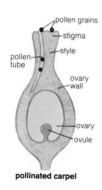

pollinated carpel

gynoecium (*n*) the whole female reproductive (p.209) organ of a flower, consisting of one or more carpels (p.211).

pistil (*n*) another name for either carpel (p.211) or gynoecium (↑). **pistillate** (*adj*).

androecium (*n*) the whole male reproductive (p.209) part of a flower, i.e. all the stamens (p.211).

receptacle (*n*) the top of a flower-stalk; its shape can vary from convex (p.59) to concave (p.59) in different flowers. The receptacle bears the perianth (↓), carpels (p.211) and stamens (p.211).

perianth (*n*) the accessory parts of a flower. It usually consists of an outer whorl (↓) of sepals (↓) and an inner whorl of petals (↓). The carpels (p.211) and stamens (p.211) lie inside the perianth.

whorl (*n*) a circle of like parts growing at the same level on the stem of a plant, e.g. a whorl of leaves on a stem, a whorl of petals on a receptacle (↑).

corolla (*n*) all the petals (↓) of a flower.

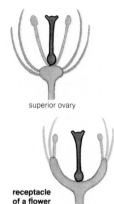

superior ovary

receptacle of a flower

superior ovary

inferior ovary

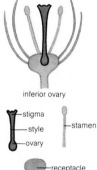

stigma
style
ovary

stamen

receptacle

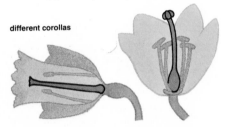

different corollas

petal (*n*) a leaf-like structure, the inner part of the perianth (↑), usually coloured to attract (p.17) insects (p.151); there are many different shapes of petals to suit the insects pollinating the flower. The petals are the flower parts most easily seen. Dicotyledons (p.155) generally have flower parts in groups of five, while monocotyledons (p.155) generally have flower parts in groups of three.

calyx (*n*) all the sepals (↓) of a flower.

sepal (*n*) in dicotyledons (p.155), a green, leaf-like structure, the outermost part of the perianth (↑). In dicotyledons (p.155) sepals are generally grouped in fives.

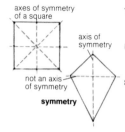

axes of symmetry of a square

axis of symmetry

not an axis of symmetry

symmetry

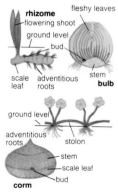

bud

yeast cell

branch

stem

buds

tentacles

body

hydra

bud

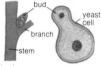

rhizome

flowering shoot

ground level

bud

fleshy leaves

scale leaf

adventitious roots

stem

bulb

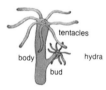

ground level

adventitious roots

stolon

stem

scale leaf

bud

corm

tepal (*n*) in monocotyledons (p.155), a structure of a perianth (↑) in which there is no difference between sepals (↑) and petals (↑).

nectary (*n*) a gland (p.162) which secretes (p.162) a sweet liquid, called **nectar**, which attracts (p.17) insects (p.151) to the flower (p.211).

symmetry (*n*) the state of having a regular shape such that a line can divide the structure into two equal parts which are similar, that is, the parts are balanced about the line. Such a line is called an *axis of symmetry*. For example, a square has symmetry as a line can be drawn through opposite corners, or through the middle of opposite sides, and in each case, the square is divided into equal parts which are balanced about the line. **symmetrical** (*adj*).

vegetative reproduction a kind of asexual reproduction (p.209) in plants. It is the growth of a new plant from a part of an old plant but not from spores (p.146), e.g. as in rhizomes (↓), bulbs (↓), corms (↓).

bud (*n*) (1) a small, pointed structure on a stem. A bud grows into either a leaf or a flower. (2) a bud-like growth from the wall of a cell which becomes large, leaves the parent cell and becomes a daughter cell, e.g. buds on yeast cells. (3) a bud-like growth from the body of certain simple animals. The bud grows into a young animal and leaves the parent, e.g. a bud on hydra, *see diagram*.

bulb (*n*) a modified (p.158), very small, under-ground stem covered in succulent (p.155) fleshy leaves which store food. Buds grow on the stem between the leaves. The new plant forms new bulbs in vegetative reproduction (↑), e.g. onion.

rhizome (*n*) an underground stem with buds (↑) in the axils (p.158) of scale-like leaves. The rhizome grows year after year, and plants grow from the buds. A way of vegetative reproduction (↑).

corm (*n*) a short, thick, round, underground stem, which stores food. The corm has buds in the axils (p.158) of scale-like leaves. A new plant grows from the bud using the stored food, and produces new corms before it dies, e.g. gladioli. A way of vegetative reproduction (↑).

stolon (*n*) a stem that grows along the ground. At a node (p.157) roots grow into the earth and a new plant grows from the node, e.g. strawberry. A way of vegetative reproduction (p.213).

runner (*n*) a stolon (↑) that roots only at its end.

tuber (*n*) the swollen end of an underground stem; it has buds in the axils (p.158) of scale-like leaves. The buds grow into new plants which produce new tubers. The tuber stores food for the new plant, e.g. potato. A way of vegetative reproduction (p.213).

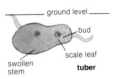

ground level

bud

scale leaf

swollen stem

tuber

vegetative propagation the use by man of parts of a plant which will grow by vegetative reproduction, e.g. cuttings (↓), grafts (↓).

cutting (*n*) a piece of a stem (p.156) put in earth; roots grow from a node (p.157) and a new plant grows from the roots by vegetative propagation (↑).

graft (*v*) (1) to produce union between the tissues of two different plants, e.g. a piece of stem or a bud is taken from one tree, and put into a cut made in the phloem (p.160) of another tree. A way of vegetative propagation (↑). (2) to produce union between the tissues of two different persons, e.g. skin graft. **graft** (*n*).

transplant (*v*) (1) to take a plant, usually a seedling (p.156), out of the earth and put it in another place. (2) to take an organ out of one animal's body and put it in the body of another animal, e.g. to transplant a kidney.

testis (*n*) (*testes*) in animals, the organ which produces sperms (↓). In vertebrates (p.148), it also produces hormones (p.208); it is a male gonad (p.210).

testicle (*n*) a testis in mammals. It is contained in a bag-like structure.

seminiferous tubule in the testes of vertebrates (p.148), a coiled tube, in man about 50 cm long and 0.2 mm in diameter, which produces sperms (↓). There are several hundred seminiferous tubules in a testis.

sperm (*n*) in animals, a male gamete (p.210); it has a nucleus (p.139), very little cytoplasm (p.138) and, in most animals, a flagellum (p.144) which allows it to swim in a liquid towards a female gamete.

spermatozoon (*n*) (*spermatozoa*) a sperm.

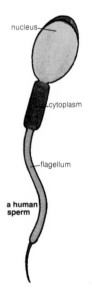

nucleus

cytoplasm

flagellum

a human sperm

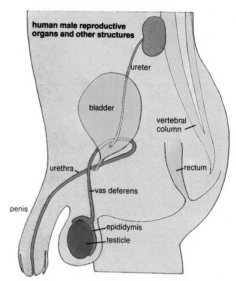

human male reproductive organs and other structures

ureter
bladder
vertebral column
urethra
rectum
vas deferens
penis
epididymis
testicle

epididymis (*n*) in reptiles (p.150), birds, and mammals (p.150), a long tube which receives sperms from a testis (↑), stores them, and then passes them to the vas deferens (↓).

vas deferens one of two tubes, one on each side of the body, passing sperm from a testis (↑), in fishes and amphibians (p.149) to a cloaca (p.165) or from an epididymis to a cloaca (p.165) in reptiles (p.150) and birds, and the urethra in mammals (p.150).

urethra (*n*) in mammals (p.150), a tube from the urinary bladder to outside the body; in males, it passes through the penis (↓). In males, the urethra is joined by the vas deferens (↑); it is the path for urine and for sperms. In females, it is a tube from the bladder alone.

penis (*n*) in mammals (p.150), an organ with many blood vessels, which is used for internal fertilization (p.210) by putting sperms into the vagina (p.217) of females. It contains erectile (↓) tissue.

erectile (*adj*) able to be stiff and upright when supplied with blood.

reproductive system the organs and structures concerned with the production of gametes.

ovary² (*n*) in animals, the organ which produces ova (↓). In vertebrates (p.148), it also produces hormones (p.208); it is a female gonad (p.210).

ovum (*n*) (*ova*) in animals, a female gamete (p.210), an unfertilized egg-cell. An ovum contains a nucleus (p.139), a lot of cytoplasm (p.138), yolk (↓) grains, and is covered by a thick membrane (p.138). In reptiles (p.150) and birds, the ovum is covered by a shell.

fallopian tube in female mammals (p.150), a tube leading from an ovary (↑) to the uterus (↓). It has a funnel-shaped opening near the ovary, and is lined with cilia (p.144) which help to conduct ova from the ovary to the uterus. Sperms (p.214) fertilize (p.210) ova in the fallopian tube.

uterus (*n*) (*uteri*) in female mammals (p.150), a hollow organ in which the embryo (↓) develops until it is born. In humans and monkeys, there is only one uterus, but in other mammals there are two, one for each fallopian tube. The walls of the uterus are formed from involuntary muscle, and the inside of the wall is covered with glandular (p.162) tissue. **uterine** (*adj*).

cervix (*n*) a short, narrow tube leading from the uterus (↑) to the vagina (↓). It secretes (p.162) mucus (p.190) to the vagina. **cervical** (*adj*).

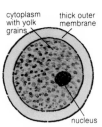

cytoplasm with yolk grains · thick outer membrane · nucleus

human ovum

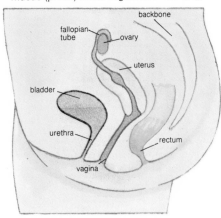

backbone · fallopian tube · ovary · uterus · bladder · urethra · rectum · vagina

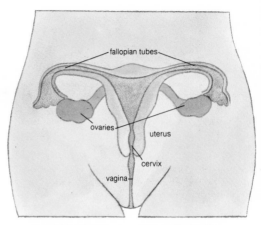

vagina (*n*) in female mammals, a tube leading from the cervix (↑) to outside the body. It receives sperm (p.214) for fertilization (p.210), and is the passage for the birth of an offspring (p.210). **vaginal** (*adj*).

egg (*n*) the ovum together with a yolk (↓) covered by a shell in birds, a tough membrane (p.138) in reptiles (p.150), or a jelly-like material in amphibians (p.149). The ovum is an *egg-cell*.

yolk (*n*) a store of food for an embryo (↓) in the eggs (↑) of most animals. It consists of protein and fat. **yolky** (*adj*).

incubate (*v*) to keep eggs under suitable conditions until the offspring (p.210) break out of the shell or membrane.

hatch (*v*) (1) of offspring, to come out of an egg, (2) of eggs, to bring forth young after incubation (↑).

embryo[2] (*n*) an animal growing in an egg or in its mother's body. Hatching (↑) ends the embryonic stage in animals other than mammals (p.150). **embryonic** (*adj*).

foetus (*n*) in mammals, the stage when an embryo (↑) starts to have the appearance of a fully developed offspring. In man, an embryo changes to a foetus after about two months. **foetal** (*adj*).

foetus in uterus

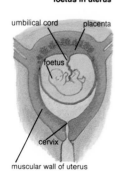

umbilical cord placenta

foetus

cervix

muscular wall of uterus

placenta (*n*) an organ of nutrition (p.171) for a foetus (p.217). It consists of foetal (p.217) tissues and mother's tissues with the two tissues interdigitated (↓). The circulatory systems of foetus and mother are quite separate. Oxygen, glucose, amino acids and fats diffuse (p.27) from the mother's capillaries (p.179) into the foetal capillaries, and carbon dioxide and urea diffuse in the opposite direction. **placental** (*adj*).

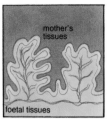
interdigitation in a placenta

interdigitate (*v*) of two tissues, to have finger-like structures of one tissue go into the hollows of the other tissue. **interdigital** (*adj*).

crypts (*n.pl.*) small hollows in the thickened wall of a uterus into which the placenta grows.

umbilical cord a tube from the placenta (↑) to the abdomen (p.162) of the foetus. It contains the foetal artery and vein. It breaks, or is broken, at birth.

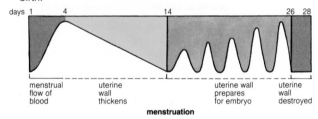

days 1	4		14		26	28

menstrual flow of blood	uterine wall thickens	uterine wall prepares for embryo	uterine wall destroyed

menstruation

menstruation (*n*) in monkeys, apes, and humans, a periodic (p.64) sending out of blood and mucus (p.190) from the uterus; in a woman, this happens every 28 days. Menstruation results from the destruction of the wall of the uterus. After menstruation, the uterine wall thickens and prepares to receive an embryo. If fertilization has not taken place, the uterine wall is destroyed, and finally sent out as blood and mucus through the vagina (p.217). Menstruation stops temporarily during pregnancy. *See diagram.* **menstruate** (*v*), **menstrual** (*adj*).

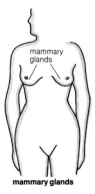
mammary glands

lactation (*n*) the production of milk in mammary glands (↓). **lactate** (*v*).

mammary glands in female mammals (p.150) glands on the abdomen (p.162) of most mammals, but on the thorax (p.189) of apes and humans.

a nucleotide

chain of nucleotides

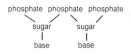

a nucleic acid

model of a double helix

each strand of DNA is a
sugar-phosphate chain

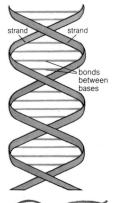

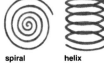

spiral   helix

nucleic acid a long chain molecule made up of a large number of nucleotides (↓). All organisms (p.147) have nucleic acids present in their cells. There are two kinds of nucleic acid, DNA (↓) and RNA (↓).

nucleotide (*n*) one part, or unit, of a nucleic acid. It consists of a sugar (p.173), either ribose, or deoxyribose, together with a phosphate radical (p.116) and a base (p.114) containing nitrogen. Each unit combines with two other units and a chain is formed.

DNA this represents deoxyribonucleic acid, a nucleic acid (↑). The sugar (p.173) is deoxyribose. The base (p.114) in each nucleotide (↑) is one of: thymine, cytosine, adenine and guanine. The long chain of nucleotides forms a **strand**, and two strands are coiled round each other to form a double helix (↓), *see diagram*. The strands are joined by bonds (p.109) between pairs of different bases, one on each strand. DNA is found only in the nuclei (p.139) of cells; DNA and protein form chromosomes. From a strand of DNA, a strand of RNA is produced by the action of an enzyme (p.167).

RNA this represents ribonucleic acid, a nucleic acid (↑). The sugar (p.173) is ribose. The base (p.114) in each nucleotide (↑) is one of: uracil, cytosine, adenine and guanine. The chain of nucleotides forms a single strand. RNA is found in the nuclei (p.139) and the cytoplasm (p.138) of cells. Strands of RNA are produced in the nucleus from DNA (↑), passed to the cytoplasm, and then a ribosome (p.141) is joined to the RNA. The ribosome moves along the strand of RNA and produces a polypeptide (p.172); the structure of the polypeptide is controlled by the RNA.

spiral (*n*) a line which starts at a point and then curves so that it gets farther and farther away from the starting point; it is a flat curve. Any structure of a similar shape. **spiral** (*adj*).

helix (*n*) a line which curves in a circle while moving away at right angles to the start of the curve. *See diagram*.

helical (*adj*) with the shape of a helix or spiral.

gene (*n*) a short length of a chromosome (p.142) which controls a characteristic (p.147) of an organism (p.147). The gene can be passed on from parent to offspring (p.210), e.g. a gene for eye-colour.

genetic (*adj*) concerned with genes (↑).

diploid (*adj*) of a nucleus, possessing chromosomes (p.142) in pairs. All cells in an organism, except gametes (p.210), have a diploid nucleus.

haploid (*adj*) of a nucleus, possessing unpaired chromosomes (p.142), i.e. half the number of chromosomes. All gametes (p.210) have a haploid nucleus. When two haploid nuclei undergo fusion (p.210), a diploid (↑) nucleus is formed, with all chromosomes paired. In man, there are 23 pairs of chromosomes in all cells except the cells of gametes, which have half the number of a diploid nucleus, i.e. 23 unpaired chromosomes.

allele (*n*) one of a pair of genes (↑) that control the same characteristic (p.147), but have a different effect, e.g. there are two genes for eye-colour, one gene producing brown eyes and the other gene producing blue eyes. These two genes are alleles of each other. Each chromosome of a pair bears a particular gene in a fixed place along its length, so that the two genes controlling a particular characteristic are in corresponding places. The two genes can be identical (p.94) or can be alleles. There can be more than two alleles for a gene, e.g. there are three alleles determining the blood groups of the ABO system (p.162) but any one pair of chromosomes has only two of these alleles.

homozygous (*adj*) of persons, possessing two identical (p.94) genes (↑) for a characteristic (p.147). See heterozygous (↓).

heterozygous (*adj*) of persons, possessing two alleles (↑) of a gene, i.e. two different genes for the same characteristic.

dominant (*n*) of an allele (↑), determines the effect of a gene, e.g. the allele for brown eyes is dominant, so a person who is heterozygous (↑) for the gene will have brown eyes, and the allele for blue eyes will have no effect.

gene

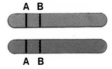

chromosome pair – homozygous for gene A and gene B

chromosome pair – homozygous for gene A, heterozygous for gene B

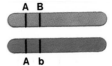

B and b are alleles of the gene

some possible combinations of 3 alleles

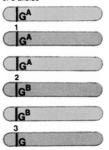

recessive (*adj*) of an allele (↑), has no genetic (↑) effect in a heterozygous (↑) person, e.g. the allele for blue eyes is recessive, so a person who is heterozygous for the gene will have brown eyes produced by the allele for brown eyes. A person with two identical genes for blue eyes will have blue eyes, as no dominant gene is present.

phenotype (*n*) a person belonging to a type which has a particular characteristic (p.147), e.g. a brown-eyed person for the characteristic of eye-colour. Such a person can be homozygous (↑) or heterozygous (↑) for a dominant (↑) gene, but must be homozygous for a recessive (↑) gene.

genotype (*n*) a person as determined by his genetic make-up, e.g. homozygous (↑) for dominant (↑) or recessive (↑) genes, or heterozygous for the gene.

X-chromosome a sex chromosome (p.142). In humans, a pair of X-chromosomes produce a female. The X-chromosome has many genes for which there are no paired genes on the Y-chromosome (↓).

Y-chromosome a sex-chromosome (p.142). It is smaller in length than an X-chromosome (↑). A nucleus (p.139) can have only one Y-chromosome, the other sex chromosome being an X-chromosome. In humans, an XY chromosome pair produces a male.

inherit (*v*) of offspring (p.210), to receive characteritistics (p.147) from parents. All characteristics of an organism are determined by the chromosomes (p.142) in the nucleus (p.139) of every cell. In a pair of chromosomes inherited by sexual reproduction, one comes from the father and one from the mother because each gamete (p.210) has a haploid (↑) nucleus, and the offspring has cells with diploid (↑) nuclei. The offspring thus inherits genes from both parents and its characteristics will be determined by the characteristics of both parents. **heritable** (*adj*), **inheritance** (*n*).

pass on (*v*) of parents, to give characteristics (p.147) to offspring (p.210).

parents both brown-eyed
both heterozygous for gene

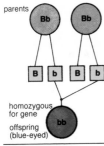

parents

homozygous
for gene

offspring
(blue-eyed)

phenotypes

brown-eyed

blue-eyed

genotypes

B allele for brown eyes -dominant

b allele for blue eyes -recessive

BB homozygous for brown eyes

bb heterozygous for brown eyes

Bb homozygous for blue eyes

inheritance of genes
to show two brown-eyed parents may produce a blue-eyed child.

Mendel's laws if in two animals, or plants, one
is homozygous (p.220) for a dominant (p.220)
gene (p.220) for the characteristic (p.147) and
the other is homozygous for a recessive (p.221)
gene for the characteristic, then the offspring
(p.210) reproduced by sexual reproduction
(p.209) will all have the characteristic from the
dominant gene. If these offspring, which are
heterozygous (p.220) for the gene, reproduce a
second generation (p.210), then the offspring of
the second generation will have 75% with the
characteristic for the dominant gene and 25%
with the characteristic for the recessive gene,
i.e. a 3:1 ratio. *See diagram.*

heredity (*n*) the passing on (p.221) of
characteristics (p.147) from one generation
(p.210) to the next generation. **hereditary** (*adj*).

pedigree (*n*) a diagram showing the inheritance
(p.221) of particular characteristics (p.147) from
one generation (p.210) to later generations.

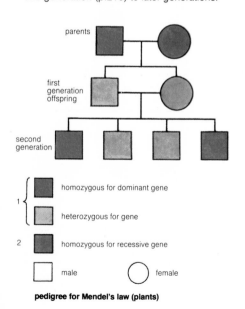

pedigree for Mendel's law (plants)

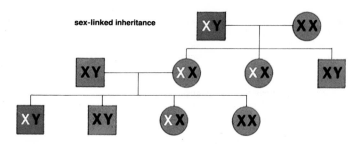

sex-linked inheritance

○ normal female

■ male with haemophilia

■ normal male

Ⓧ recessive gene for haemophilia

X dominant normal gene

Y no gene on Y chromosome

sex-linked of a gene, carried on the X-chromosome (p.221) but not on the Y-chromosome (p.221). A man has only one X-chromosome, so that men have the characteristics from recessive (p.221) genes (p.220) in sex linkage more often than women, who have two X-chromosomes. A son cannot inherit (p.221) a sex-linked gene from his father, but his daughter can inherit the gene on the X-chromosome. The daughter in turn can carry or pass on (p.221) the gene to her son, so a recessive gene can appear in male offspring (p.210) every other generation. Colour-blindness is a sex-linked gene. **sex linkage** (*adj*).

haemophilia (*n*) a disease of humans in which blood does not clot properly. The sex linkage (↑) of the gene makes it common in men, but uncommon in women. Unless a woman has two X-chromosomes with the recessive (p.221) gene she will not suffer from haemophilia; with only one recessive gene she is a carrier.

birth rate the number of children born in a year per thousand people, e.g. a birth rate of 38 per 1000 means 38 000 children born yearly in a population (p.224) of 1 000 000.

death rate the number of people dying in a year per thousand people, e.g. a death rate of 22 per 1000 means 22 000 people die in a population (p.224) of 1 000 000.

mortality rate the number of deaths in a certain period per 1000 people in an age group; usually the age group is for a five-year period, e.g. the mortality rate for the age group 30–34 years.

population (*n*) the number of organisms (p.147) in a given place or area, e.g. (a) the number of lions in Africa; (b) the number of date palms in Egypt; (c) the mosquito population of a lake. Bacteria and protozoa are not usually considered as a population. If no particular organism is named, the population is the number of persons living in a named place. **populate** (*v*).

survival (*n*) the act of living through unfavourable conditions until the conditions change. **survive** (*v*).

evolution (*n*) the changes over many generations (p.210) by which different kinds of organisms (p.147) have arisen from very early forms, e.g. about 400 million years ago a reptile, called a pterodactyl, flew in the air. About 200 million years ago, many changes in previous generations resulted in an early kind of bird, called an Archaeopteryx. Many more changes resulted in the present kind of birds, which appeared in the last 10 million years or so. These changes have been studied from fossils (↓) and show the evolution of birds from reptiles.

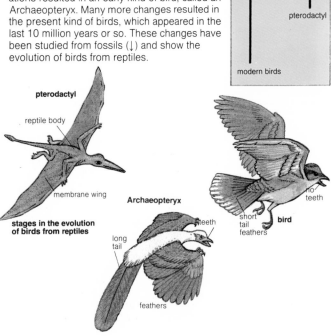

evolution

early forms

Archaeopteryx

pterodactyl

modern birds

pterodactyl

reptile body

membrane wing

stages in the evolution of birds from reptiles

Archaeopteryx

teeth

long tail

feathers

short tail feathers

no teeth

bird

fossil (*n*) the remains of an organism (p.147), or the shape of an organism, preserved in rocks. The hard parts of the organism are generally changed into rock by mineral salts dissolved in the water passing through the earth. The organisms that lived many millions of years ago can be studied from their fossils.

fossil

mutation (*n*) a sudden change in the genes (p.220) of an organism (p.147) caused by a change in the DNA (p.219) of the chromosomes (p.142). Mutation does not happen often, but the rate is increased by radiation (p.83), neutrons (p.106) and some chemical substances. Changes of single genes can take place, or a whole chromosome can be altered. Changes in the genes in body cells have an effect only on the person; but changes in the genes in gametes (p.210) have an effect on all the offspring (p.210). All mutations happen by chance. Most mutations have a bad effect. **mutate** (*v*).

mutant (*n*) (1) a gene (p.220) which has been altered by mutation (↑). (2) an organism (p.147) altered by such a gene. (3) a characteristic (p.147) produced by such a gene.

natural selection the tendency (p.15) to survive (↑) of those organisms (p.147) which are best suited to their conditions of living. These organisms live longer and reproduce (p.209) offspring (p.210) with inherited (p.221) characteristics (p.147) which are useful for survival (↑). The organisms and offspring which survive are said to be selected by a natural process. Natural selection thus controls the direction of change of inherited characteristics as unsuitable mutants (↑) die.

Darwinism (*n*) the idea that evolution (↑) took place by natural selection (↑).

biosphere (*n*) the part of the Earth and the
    atmosphere (p.51) in which all organisms
    (p.147) live.
community (*n*) a group of organisms (p.147)
    living in a certain habitat (↓), having an effect on
    each other, and reaching a state of equilibrium
    (p.20) through a food web (p.171).
biome (*n*) a major community (↑) covering a large
    area of the Earth, e.g. tundra (very cold areas
    with no trees, and low plant growth); tropical
    rain forest; savanna (grass lands); desert.
environment (*n*) all the conditions which act on
    an organism (p.147) while it lives. Besides the
    physical conditions, there are also the effects of
    other organisms. The major environments are
    the sea, the fresh waters, and the land. These are
    divided into smaller environments by the
    conditions of climate (p.51). **environmental**
    (*adj*).
habitat (*n*) the place, or kind of place, where a
    particular organism (p.147) can survive (p.224),
    e.g. a sea-shore. Various factors (↓) determine
    whether a place can be a habitat for an organism,
    e.g. the habitat of the mangrove tree is a
    tropical area with brackish (↓) water
    over a bottom of mud.
brackish (*adj*) describes water less
    salty than sea water.

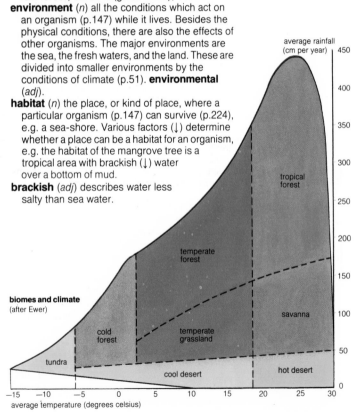

biomes and climate
(after Ewer)

biotope (*n*) an area in which animals and plants having similar habits and needs live together and form a community (↑). Also a very small habitat (↑), e.g. a pond, a tree.

biocoenosis (*n*) a community (↑) living in a biotope (↑).

ecosystem (*n*) a complete ecological (↓) unit which can be studied. It is a community (↑) which acts on the environment (↑) and the environment acts on the community. In a biotope (↑), the biocoenosis (↑) acts on the biotope, and the biotope acts on the biocoenosis. An ecosystem consists of (a) *producers*, mainly green plants; (b) *consumers*, animals feeding on plants and other animals; (c) *decomposers*, mainly bacteria (p.145) and fungi (p.145) decomposing dead organisms.

territory (*n*) the area in which an animal, or a society (p.153) of animals, lives, e.g. ants from a particular ant hill cover an area for feeding and fight other ants trying to enter their territory. A single bird can have a territory for feeding or mating. Sometimes a territory is held for only part of a life-cycle (p.151). **territorial** (*adj*).

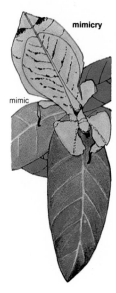

mimicry

mimic

factor (*n*) a factor helps to determine an environment (↑), e.g. climate (p.51), the kind of soil, salinity (p.228) of water, the presence of organisms (p.147) are some of the factors determining an environment. See conditions (p.228).

adaptation (*n*) a change in characteristics (p.147) of an organism (p.147) which increases its chances of survival (p.224). Such changes can come through evolution (p.224), or can be physiological (p.140) or sensory (p.200), e.g. fish living in deep water have eyes stimulated best by blue light, the only light at that depth; this is evolutionary adaptation. **adapt** (*v*).

mimicry (*n*) of an animal, being the same in appearance as another animal, or a plant, e.g. (a) a stick insect looks like a part of a plant; (b) an animal having coloured marks on it so it looks like a poisonous animal. **mimic** (*v*).

ecology (*n*) a study of the relations between animals, plants and the physical conditions of the environment; in particular the study of ecosystems (↑). **ecological** (*adj*).

conditions (*n.pl.*) the actual measurable physical factors (p.227) in an environment (p.226), e.g. temperature is a physical factor of climate (p.51) but the actual range (p.108) of temperature in a particular area is a condition.

salinity (*n*) a measure of the amount of common salt in water. The average salinity of sea water is about 2.8 g sodium chloride per 100 g water. **saline** (*adj*).

aquatic (*adj*) living in water; concerned with water.

freshwater (*adj*) living in water which contains no salt; concerned with water which contains no salt, e.g. rivers and lakes are freshwater environments.

sea-water (*n*) water in the seas usually containing 2.8% sodium chloride, 0.4% magnesium chloride, 0.2% magnesium sulphate, 0.1% calcium sulphate, 0.1% potassium chloride.

marine (*adj*) living in sea-water; concerned with the sea, e.g. a marine plant; marine ecology.

estuarine (*adj*) living in an estuary (↓); concerned with estuaries.

aquatic organisms

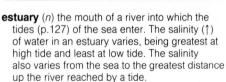

diagram of an estuary

estuary (*n*) the mouth of a river into which the tides (p.127) of the sea enter. The salinity (↑) of water in an estuary varies, being greatest at high tide and least at low tide. The salinity also varies from the sea to the greatest distance up the river reached by a tide.

littoral (*adj*) living on or near a sea-shore; living in a lake near the shore on the lake bottom, when the lake bottom is at a depth of less than 10 m. The littoral zone of a sea is the part of the sea between high and low tide marks.

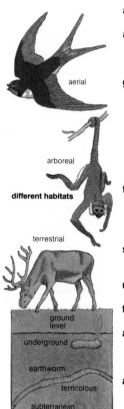

aerial

arboreal

different habitats

terrestrial

ground
level

underground

earthworm

terricolous

subterranean

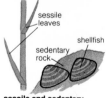

sessile
leaves

shellfish

sedentary
rock

**sessile and sedentary
organisms**

amphibious (*adj*) living both on land and in water, e.g. frogs are amphibious animals.

amphibiotic (*adj*) living in water for the first stages (p.152) of a life cycle (p.151) and on land, when fully grown, e.g. mosquitoes are amphibiotic insects.

ground (*n*) the surface of the Earth; the earth on which animals move and plants grow; the support for a structure, e.g. (a) ground speed is the speed of an aeroplane in relation to the Earth; (b) ground level is the height above sea-level of the Earth's surface at a particular point; (c) ground substance is the supporting material between cells, e.g. a matrix (p.192) of fibres (p.195). **ground** (*adj*).

terrestrial (*adj*) living on the ground (↑), concerned with the Earth, e.g. (a) terrestrial magnetism is the magnetism (p.69) of the Earth; (b) a terrestrial animal is one that walks on the ground, e.g. a deer lives in a terrestrial habitat.

subterranean (*adj*) deep down in the Earth, e.g. subterranean rock strata of metamorphic rock (p.121).

underground (*adj*) in the earth, but not deep, e.g. a rhizome (p.213) is an underground stem.

terricolous (*adj*) of animals, living in the soil (p.230), e.g. an earthworm is a terricolous organism.

arboreal (*adj*) of animals, living in trees, e.g. monkeys are arboreal animals. Also describes a habitat (p.226), e.g. monkeys live in an arboreal habitat.

aerial (*adj*) concerned with the air, living in the air, e.g. (a) the aerial roots of a plant grow above ground (↑); (b) birds have an aerial territory (p.227) for feeding.

sedentary (*adj*) of animals, living fixed to a support, not free to move from place to place, e.g. some shellfish are fixed to rocks on the sea-shore.

mobile (*adj*) of animals, free to move from place to place, capable of locomotion (p.194).

sessile (*adj*) (1) of plants and structures, fixed to a stem or support without a stalk (p.157), e.g. a sessile leaf has no petiole (p.158). (2) of animals, sedentary (↑).

motile (*adj*) of protozoa, able to move.

diurnal (*adj*) (1) active only in the day-time, during the hours of light, e.g. a diurnal animal hunts its food during daylight; (2) happening every day; some flowers have a diurnal rhythm, that is, they have changes that happen every twenty-four hours.

nocturnal (*adj*) active only at night, e.g. a nocturnal animal hunts its food at night.

soil (*n*) when earth is considered as an environment (p.226) providing nutrients (p.171) for plants and animals. Soil consists of grains which have water and air between them. The grain structure of soil is important for its properties. The properties of soil determine which kinds of plants can grow in it. Soil is formed by the weathering (p.120) of rocks (p.119) producing particles (p.26) which are mixed with organic (p.131) material.

soil profile the layers (p.187) of different kinds of soil that can be seen if a hole is dug down to the rock beneath.

topsoil (*n*) the top layer (p.187) of soil. It contains the nutrients (p.171) needed by plants. If the topsoil is taken away by erosion (p.120), plants can no longer grow in the soil left behind.

subsoil (*n*) the layers of soil beneath the topsoil (↑). Although a subsoil does not provide sufficient nutrients (p.171) for plants, it can be useful by not allowing water to drain (↓) away.

clay (*n*) soil (↑) with particles (p.26) less than 0.01 mm in diameter, mainly aluminium silicate. Water passes through it very slowly.

sand (*n*) soil (↑) with particles (p.26) between 0.1 mm and 2 mm in diameter, mainly silicon dioxide. Water passes through it very quickly. **sandy** (*adj*).

loam (*n*) a kind of soil (↑) consisting of sand and clay, the best soil for the growth of plants.

humus (*n*) organic (p.131) material produced by the decay (p.146) of plant and animal tissues; it makes soil dark in colour. It is the most important constituent (p.96) of soil (↑) for plant growth, and also helps soil to hold water.

drain (*v*) of water, to pass through soil, or to run off a surface.

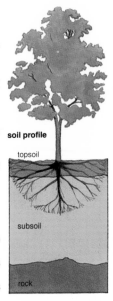

soil profile

topsoil

subsoil

rock

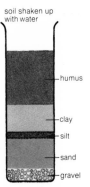

soil shaken up with water

humus

clay

silt

sand

gravel

the constituents of loam

manure (*n*) the faeces (p.169) and excreta (p.186) of animals when put on soil (↑). It helps to form humus (↑) and provides nutrients (p.171) for plants.

fertilizer (*n*) any substance which is added to the soil (↑) to provide nutrients (p.171) for plants. Manure (↑) and compost (↓) are natural fertilizers. Certain chemical substances are artificial fertilizers, e.g. ammonium sulphate is an artificial fertilizer which is changed to nitrates (p.118) for plants.

compost (*n*) decayed (p.146) plant material put on soil (↑) to provide nutrients (p.171) for plants.

agriculture (*n*) the processes of growing plants and keeping animals to obtain food and other products. To help in these processes, soil conservation (p.44) is needed so that nutrients (p.171) taken out of the soil by plants are put back by the farmer. Erosion (p.120) must also be prevented. **agricultural** (*adj*).

crop (*n*) any kind of plant, grown by agricultural (↑) methods, for food or for other products, e.g. rubber trees grown for rubber. Also the product produced, e.g. rubber is a crop.

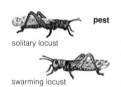

pest

solitary locust

swarming locust

pest (*n*) any organism, usually an insect, causing damage to crops (↑).

weed (*n*) a plant which is not wanted, growing in a crop (↑).

escape (*n*) a plant that was grown as a crop (↑) and is found growing wild. The seeds are dispersed (p.155) by wind or animals. The escape is not wanted and is like a weed.

edaphic factors the factors (p.227) in an environment (p.226) which are determined by the characteristics (p.147) of the soil (↑). They arise from the physical and chemical properties of the soil.

biotic factors the factors (p.227) in an environment (p.226) which arise from the activities of organisms (p.147).

climatic factors the factors (p.227) in an environment (p.226) which arise from the climate. The edaphic (↑), biotic (↑) and climatic factors form and determine the environment in a particular area.

cycle (*n*) changes and events which have no
beginning and no end, but follow in turn, so that
they are repeated continuously. All processes in
nature (p.147) must follow a cycle, otherwise a
natural process would come to an end.

carbon cycle all organic (p.131) material contains
the element (p.103), carbon, and all organisms
(p.147) also consist of carbon. A balance (p.18)
is needed between carbon as carbon dioxide in
the atmosphere (p.51) and carbon in organisms.
Respiration (p.191) and decay (p.146) produce
carbon dioxide. Green plants use carbon dioxide
from the atmosphere to synthesize (p.136) carbohydrates (p.173) used in internal respiration to
provide energy for life. The carbon cycle is shown
in the diagram below.

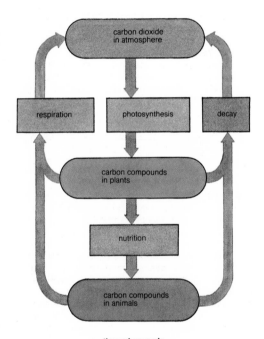

the carbon cycle

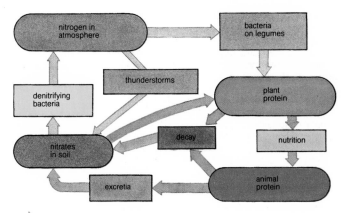

main cycle

other cycles

the nitrogen cycle

nitrogen cycle the element (p.103), nitrogen, is important as it is a constituent (p.96) of all proteins (p.172). Nitrates (p.118) in the soil (p.230) are used by plants to make proteins. Plant proteins are used by animals to produce the proteins they need for living. The nitrogen in protein is returned to the soil in animal excretion (p.186) as urine (p.186) or by the decay (p.146) of animal and plant tissues. Bacterial (p.145) caused decay produces nitrates from ammonia (p.186) set free from urea (p.186) and protein. The full cycle is shown in the diagram.

wilt (v) of plants, to lose turgor (p.139) so that the plant no longer stands upright. Wilting for a long time results in death.

succession (n) of plants, a change, with time, of the species in a community (p.226). It begins with the start of plant life, often single-cell plants, and grows towards a climax (↓).

sere (n) a plant succession (↑) in a particular kind of environment, e.g. a plant succession in water is a **hydrosere**; a plant succession in a dry environment, such as a desert, is a **xerosere**; a sere on uncovered rock is a **lithosere**.

climax (n) a plant community (p.226) at the end of a succession (↑); it is in equilibrium (p.20) with the environment (p.226) and there is no further change unless there is a change in the climate.

commensalism (*n*) the state in which animals of two different species (p.148) share the same living place and also the same food, e.g. a species of marine worm makes a hole in the sand; a small shrimp, also lives in the hole; both use the same food, but they have no effect on each other. The two animals live in a state of commensalism. **commensal** (*adj*).

symbiosis (*n*) the state in which organisms (p.147) of two different species (p.148), even a plant and an animal, live together and are useful to each other, e.g. (a) bacteria living in the roots of leguminous (p.155) plants provide nitrates for the plant, while the plant provides carbohydrates (p.173) and other food materials for the bacteria; (b) cattle carry ticks (arthropod parasites (↓)); some species of birds live on the cattle and eat the ticks. The cattle are useful to the birds as they provide food; the birds are useful to the cattle as they free them from ticks. The cattle and the birds live in a state of symbiosis. **symbiotic** (*adj*).

symbiont (*n*) one of the two organisms (p.147) living in symbiosis (↑).

mutualism (*n*) another name for symbiosis (↑).

parasite (*n*) an organism (p.147) living on, or in, another organism and obtaining all its food from its host (↓). Parasites may annoy or harm their host. Examples of parasites are: amoeba (p.143) living in the human gut (p.162); fleas living on human beings, dogs and cats. **parasitic** (*adj*).

host (*n*) a living organism (p.147) on which, or in which, a parasite (↑) lives.

saprophyte (*n*) an organism (p.147) which feeds on dead or decaying organisms. Saprophytes are mainly bacteria and fungi; they are very important as they complete the carbon and nitrogen cycles (p.233). **saprophytic** (*adj*).

epiphyte (*n*) a plant growing on another plant and using it only for support, i.e. it is not a parasite. Examples of epiphytes are ferns growing in the axils (p.158) of branches of trees. **epiphytic** (*adj*).

predator (*n*) an animal which obtains food by hunting other animals. **predatory** (*adj*).

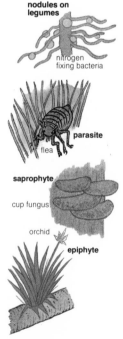

nodules on legumes

nitrogen fixing bacteria

parasite

flea

saprophyte

cup fungus

orchid

epiphyte

prey (*n*) the animals attacked and eaten by a predator (↑).

hibernation (*n*) in many mammals (p.150), most reptiles (p.150) and amphibians (p.149) a state of physiological resting, during winter. The rate of metabolism (p.172) is very low, and the temperature of mammals drops to that of the atmosphere.

herbivore (*n*) an animal obtaining all its food from plants. **herbivorous** (*adj*).

carnivore (*n*) an animal obtaining all its food as meat from other animals.
carnivorous (*adj*).

omnivore (*n*) an animal obtaining its food from both plants and other animals.
omnivorous (*adj*).

plankton (*n*) very small plants and animals living in water; they float (p.36) in the water. Most of them live near the surface of the water where the plants get sufficient light for photosynthesis (p.159). Plankton are important in aquatic (p.228) food webs (p.171) as they provide food for fishes and for animals as large as whales.

phytoplankton (*n*) plants of plankton (↑), the main producers (p.236) in an aquatic (p.228) food web. They are not capable of locomotion (p.194), so they float (or drift) in the waters of seas and lakes. They are a source of food for fishes and zooplankton (↓).

zooplankton (*n*) animals of plankton (↑). They feed on phytoplankton (↑) and most are capable of locomotion by flagellae (p.144). They provide food for fishes.

nekton (*n*) the name given to all animals that swim in seas or fresh water. Fishes, crabs, octopus, are part of nekton.

benthon (*n*) animals and plants which live on the bottom of a sea or lake. They may be sessile (p.229) on the bottom, or they may move about on the bottom. The bottom stretches from high tide (p.127) mark to the greatest depths of water.

pelagic (*adj*) living in the open waters of a sea or lake as opposed to living on the bottom of the sea or lake. Plankton (↑) and nekton (↑) are pelagic organisms.

producers (*n.pl.*) green plants that use inorganic (p.116) materials to produce carbohydrates (p.173), proteins (p.172) and fats (p.175). Energy is obtained from sunlight through photosynthesis (p.159).

consumers (*n.pl.*) animals and plants without chlorophyll (p.159) that obtain their food from other organisms. Primary consumers are herbivores (p.235), secondary consumers are carnivores (p.235). Higher order consumers are large carnivores feeding on smaller carnivores.

decomposers (*n.pl.*) bacteria (p.145), fungi (p.145), and some protozoa (p.143) which cause the decay (p.146) of dead organisms (p.147) into inorganic (p.116) materials so the carbon, nitrogen, and other cycles can be completed.

trophic level the level of producer (↑) or consumer (↑) in a food chain (p.171). The first level contains producers; all other levels are consumers. Herbivores are primary consumers, smaller and larger carnivores are secondary and tertiary consumers. The energy available from carbohydrate in grass for farm animals is shown in the diagram above.

pyramid of numbers the number of organisms at each trophic level in a food web (p.171). At each level, energy is lost from organisms through respiration, heat radiation (p.45) and other metabolic processes (p.172). The energy left is passed on to the next trophic level. At each ascending level, less energy is available; the organisms are also larger, so for both reasons there are fewer organisms at a higher trophic level than at a lower level. *See diagram.*

pathogen (*n*) a bacterium (p.145), virus (p.145), fungus (p.145) or protozoan (p.143) causing disease. Such an organism (p.147) is called an agent of disease; it is a parasite (p.234).

infect (*v*) of pathogens, to enter an organism (p.147) and cause a condition of disease, e.g. a particular virus infects a man and he suffers from yellow fever. **infection** (*n*), **infectious** (*adj*).

infest (*v*) of parasites, to live in large numbers on an animal or plant. Parasites can also infest a room or clothes. **infestation** (*n*), **infested** (*adj*).

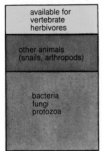

1×10^7 kilocalories per year

energy from 4000 m² grass.

energy available from grass

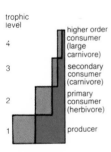

trophic level

4 — higher order consumer (large carnivore)

3 — secondary consumer (carnivore)

2 — primary consumer (herbivore)

1 — producer

energy lost through respiration, heat radiation and other metabolic processes

energy available as food

pyramid of available energy at the trophic levels of a food web

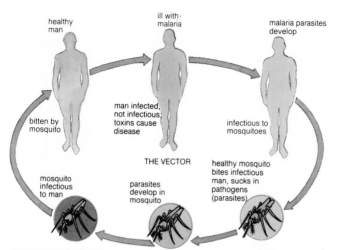

healthy man

ill with malaria

malaria parasites develop

bitten by mosquito

man infected, not infectious; toxins cause disease

infectious to mosquitoes

THE VECTOR

mosquito infectious to man

parasites develop in mosquito

healthy mosquito bites infectious man, sucks in pathogens (parasites)

a vector transmits disease

vector² (*n*) an animal or a physical way of carrying a pathogen (↑) from an infected (↑) person to a new host (p.234), e.g. (a) a mosquito is a vector for a protozoan which is the causative agent (↓) of malaria; (b) contaminated (p.240) water is the physical vector for cholera, the causative agent of which is a bacterium (p.145).

causative agent the particular pathogen which causes a disease, e.g. *Plasmodium* is a protozoan which causes malaria; it is the causative agent of the disease.

toxin (*n*) any poison produced by a plant or animal, particularly by a bacterium. **toxic** (*adj*).

venom (*n*) any poison produced by an animal, and used when attacking other animals to poison them. **venomous** (*adj*).

antigen (*n*) a foreign substance, usually a protein or carbohydrate, which enters a vertebrate (p.148) body and stimulates a chemical reaction by the body. When a pathogen (↑) or its toxin (↑) enters the tissues, the animal produces antibodies (p.238). Some pathogens have the same antigen, while a particular pathogen may bear several antigens. Blood is an example of a material bearing an antigen (see **compatible**, p.184).

antibody (*n*) a chemical substance produced by lymphocytes, and other structures in the body of a vertebrate (p.147), when an antigen (p.237) enters the animal's tissues. The antibody combines chemically with the antigen and makes it harmless. An antibody generally combines with only one particular antigen. Cell walls of bacteria (p.145) bear antigens, and antibodies are formed to make the bacteria harmless; some antibodies make bacteria more easily attacked by phagocytes (p.183). Antibodies are carried round the animal's body by blood and lymph (p.182).

antitoxin (*n*) an antibody (↑) which combines with a toxin (p.237) to make it harmless.

vaccine (*n*) a liquid containing weakened or dead pathogens (p.236). When put in the body of a vertebrate (p.148), it causes antibodies (↑) to be produced, i.e. the same antibodies as combine with the antigens (p.237) of the pathogen.

antiserum (*n*) a liquid containing antibodies (↑); the liquid is serum (p.180) taken from an animal producing the antibodies.

inoculation (*n*) the putting of a vaccine or an antiserum into the blood of a vertebrate (p.148) animal. **inoculate** (*v*).

allergy (*n*) a very strong reaction by the body of an animal to a particular antigen, e.g. some people have an allergy to pollen (p.211); it causes the nose to produce large quantities of mucus (p.190). **allergic** (*adj*).

immunity (*n*) the ability of a plant or animal to resist the attack of antigens (p.237), or pathogens (p.236). The animal defends itself by: (a) its skin not allowing pathogens to enter; (b) the acid in its stomach destroying pathogens; (c) phagocytes (p.183) digesting pathogens; (d) antibodies making antigens harmless. Immunity is needed against each particular disease or each particular antigen, so an animal can be immune to some diseases and not to others. **immune** (*adj*).

immunize (*v*) to give a person, or animal, an inoculation (↑) of vaccine or serum which provides artificial immunity (↑) against a particular disease. **immunization** (*n*).

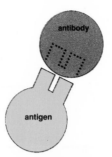

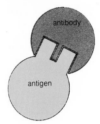

action of antibody and antigen

acquired immunity this kind of immunity (↑) is obtained by a person when he suffers from a disease and gets well again. His blood contains antibodies (↑) against the pathogens (p.236) or toxins (p.237), which caused the disease, so that his body resists further attacks. The immunity is for a particular disease and the length of time for which he is protected varies with the disease, e.g. immunity against measles is for life, that against influenza is only for a short period.

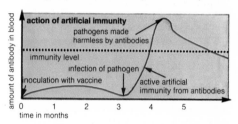

artificial immunity this kind of immunity (↑) is obtained by an inoculation (↑) and only lasts for a short time, e.g. between six months and six years, depending on the disease.

passive artificial immunity immunity obtained from an antiserum (↑). It is given to a person suffering from a disease to make him well.

active artificial immunity immunity (↑) obtained from a vaccine (↑). It is given to a healthy person, and prevents his being attacked by a particular disease by stimulating production of antibodies.

epidemic (*n*) a state of affairs in which a great many people are attacked by the same disease at the same time, so that many suffer from the disease. **epidemic** (*adj*).

endemic (*adj*) of diseases, always present in particular places or particular groups of people.

pandemic (*adj*) of diseases, spreading over the world.

sporadic (*adj*) of diseases, present from time to time in particular places, happening irregularly.

isolate (*v*) to keep a person separated from all other people so that he cannot infect (p.236) them.

hygiene (*n*) the science of keeping healthy by cleanliness, proper diet, enough exercise, and not eating contaminated (↓) food or using contaminated water. **hygienic** (*adj*).

contaminate (*v*) to spread viruses (p.145), bacteria (p.145) or other pathogens (p.236) on food, water, clothes, or in the air, e.g. (a) houseflies, can contaminate food when they walk on it; (b) faeces from diseased persons can contaminate water.

pollute (*v*) to spread harmful or unpleasant substances in air, water or on the ground, e.g. (a) the smoke from fires pollutes the air; (b) animal faeces can pollute river water; (c) sea shores may be polluted by oil from ships. **pollution** (*n*).

antiseptic (*adj*) (1) describes chemical substances used on cuts and wounds to prevent pathogens (p.236) entering. (2) of conditions, preventing pathogens entering cuts and wounds. **antiseptic** (*n*).

aseptic (*adj*) of conditions, with no pathogens (p.236) present.

antibiotic (*n*) a substance which prevents bacteria reproducing; this allows the body's phagocytes (p.183) to destroy the bacteria. Antibiotics have no effect on most viruses (p.145), protozoa (p.143) or fungi (p.145). **antibiotic** (*adj*).

pasteurization (*n*) a process (p.129) of making liquids and food free from bacteria (p.145). The food or liquid is heated to a temperature at which the bacteria are killed, but the food or liquid is not spoilt.

anaemia (*n*) a condition in which the blood has too few red blood cells (p.179) or too little haemoglobin (p.179) in the red blood cells. The blood carries too little oxygen to the tissues, so a person has little energy. **anaemic** (*adj*).

leukaemia (*n*) a condition in which a person's blood contains too many white blood cells (p.180). The spleen (p.183) becomes very large and the person usually dies.

cancer (*n*) a growth of abnormal cells in epithelial (p.192) tissue. The growth increases with time and eventually may cause death. **cancerous** (*adj*).

pollution

smoke

factory

river   effluent

atrophy (*n*) a decrease in size of an organ, or part of the body; usually caused by lack of use. **atrophy** (*v*).

infestation (*n*) the condition of being a host to parasites (p.234), other than pathogens (p.236), e.g. an infestation of fleas.

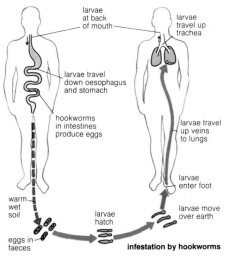

larvae at back of mouth

larvae travel up trachea

larvae travel down oesophagus and stomach

hookworms in intestines produce eggs

larvae travel up veins to lungs

larvae enter foot

warm wet soil

larvae move over earth

eggs in faeces

larvae hatch

infestation by hookworms

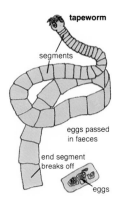

tapeworm

segments

eggs passed in faeces

end segment breaks off

eggs

tapeworm (*n*) a parasite (p.234) in the gut (p.162) of vertebrates (p.148). Its body is a chain of segments, each one complete in itself, *see diagram*. Each segment has both male and female gonads (p.210). The eggs of a tapeworm are passed out in the host's faeces.

hookworm (*n*) a parasite (p.234) in the gut (p.162) of man. Male and female hookworms fix themselves to the walls of the intestine and feed on the blood of their host. The male fertilizes (p.210) the female, and she produces eggs which are passed out in faeces. Larvae (p.151) hatch out of the eggs, grow, change into mobile (p.229) larvae, which enter the feet of human beings. Fully-grown hookworms in the gut are 2 to 3 cm long; they cause anaemia (↑) and general ill-health.